# Biodiversity Research Methods

# Biodiversity Research Methods

## IBOY in Western Pacific and Asia

Kyoto University Press

First published in 2002 jointly by:

Kyoto University Press
Kyodai Kaikan
15-9 Yoshida Kawara-cho
Sakyo-ku, Kyoto 606-8305, Japan
Telephone: +81-75-761-6182
Fax: +81-75-761-6190
Email: sales@kyoto-up.gr.jp
Web: http://www.kyoto-up.gr.jp

Trans Pacific Press
PO Box 120, Rosanna, Melbourne
Victoria 3084, Australia
Telephone: +61 3 9459 3021
Fax: +61 3 9457 5923
E-mail: enquiries@transpacificpress.com
Website: http://www.transpacificpress.com

Copyright © Kyoto University Press and Trans Pacific Press 2002

Set by digital environs, Melbourne: enquiries@digitalenvirons.com

Printed in Melbourne by McPherson's Printing Group, Maryborough, Victoria

**Distributors**

*Australia*
Bushbooks
PO Box 1958, Gosford, NSW 2250
Telephone: (02) 4323-3274
Fax: (02) 9212-2468
Email: bushbook@ozemail.com.au

*UK and Europe*
Drake International Services
PO Box 733, Cardiff CF14 2YX
UK
Telephone: (0292) 056-0343
Fax: (0292) 056-1631
Email: info@drakeint.co.uk
Web: http://www.drakeint.co.uk

*USA and Canada*
International Specialized Book Services (ISBS)
5824 N. E. Hassalo Street
Portland, Oregon 97213-3644
USA
Telephone: (800) 944-6190
Fax: (503) 280-8832
Email: orders@isbs.com
Web: http://www.isbs.com

The publication of this volume was made possible with the generous support of the Kyoto University Foundation.

All rights reserved. No production of any part of this book may take place without the written permission of Kyoto University Press or Trans Pacific Press.

ISBN  4–87698–435–2 (Kyoto University Press)
ISBN  1–87684–377–2 (Trans Pacific Press)

**National Library of Australia Cataloging in Publication Data**

Biodiversity research methods : IBOY in Western Pacific and Asia.

Bibliography.
Includes index.
ISBN 1 87684 377 2 (Trans Pacific).

ISBN 4 87698 435 2 (Kyoto University).
1. Biological diversity – Research – Methodology.
2. Biological diversity – Asia. 3. Biological diversity – Pacific Area. I. Stork, N. E. II. Nakashizuka, T. (Tohru), 1956–.

333.95

# Contents

*List of figures* xii
*List of tables* xv
*Preface* xvi

**Chapter 1: What is IBOY** 1
1.1 Mission for IBOY-DIWPA 1
  1.1.1 What is IBOY-DIWPA? 1
  1.1.2 What are the goals of IBOY-DIWPA? 1
  1.1.3 Strategy for implementing IBOY-DIWPA 5
  1.1.4 Integrated approaches to biodiversity assessment and capacity building 5
  1.1.5 Study sites 7
  1.1.6 Selection of target organisms 7
1.2 Mission for biodiversity assessment 9
  1.2.1 Why is biodiversity assessment essential? 9
  1.2.2 How frequently and where should biodiversity assessments be made? 12
  1.2.3 At what biodiversity level should assessments be made? 12
    *At the genetic level* 12
    *At the population level* 14
    *At the species level* 15
    *Species turnover in ecosystems* 16
    *At the ecosystem and landscape levels* 17
  1.2.4 The rationale for inventorying and monitoring inside and outside protected areas 18
    *Status of identified features in protected areas* 19
    *Threats to protected areas* 19
    *Use and socio-economic benefits of protected areas* 20
    *Inventorying and monitoring outside protected areas* 20
  1.2.5 Scales, planning and approaches to inventorying and monitoring 21
References 23

**Chapter 2: Forest Ecosystems** 27
2.1 Introduction 27
  2.1.1 Forest ecosystems and IBOY 27

| | |
|---|---|
| 2.1.2 Secondary transects | 28 |
| 2.1.3 General rationale and the goals of surveys | 28 |
| 2.1.4 Core and satellite sites | 30 |
| 2.2 Selection and establishment of observation sites | 32 |
|   2.2.1 Site selection | 32 |
|     *Accessibility* | 32 |
|     *Adjacency to laboratory facilities* | 33 |
|     *Environmental uniformity* | 33 |
|     *Topography* | 33 |
|   2.2.2 Plot establishment | 33 |
| 2.3 Environmental variables | 36 |
|   2.3.1 Meteorological data | 36 |
|   2.3.2 Light environment | 36 |
| 2.4 Plant diversity | 38 |
|   2.4.1 Tall-trees | 38 |
|     *Targets* | 38 |
|     *Parameters of forest dynamics* | 38 |
|     *Diversity* | 39 |
|   2.4.2 Other components | 39 |
|     *Targets* | 39 |
|     *Diversity* | 39 |
|   2.4.3 Litter supply and decomposition rate | 39 |
|     *Supply* | 39 |
|     *Accumulation* | 42 |
|     *Rate of decomposition* | 42 |
|   2.4.4 Phenology | 42 |
| 2.5 Animal diversity: arthropods | 42 |
|   2.5.1 General trapping methods | 43 |
|     *Light traps* | 43 |
|     *Malaise traps* | 48 |
|     *Window traps* | 52 |
|     *Canopy knockdown* | 53 |
|     *Bark spraying* | 58 |
|     *Pitfall traps* | 59 |
|     *Leaf litter sampling* | 63 |
|     *Soil sampling* | 65 |
|   2.5.2 Surveys of selected taxa | 67 |
|     *Drosophilidae* | 67 |
|     *Lepidoptera* | 69 |

|  |  |
|---|---|
| *Spiders* | 73 |
| *Ants* | 75 |
| *Termites* | 80 |
| 2.5.3 Handling, identification, storage and data management for arthropod surveys | 87 |
| *Sorting and identification* | 89 |
| *Storage* | 91 |
| *Data storage and handling* | 91 |
| *Backups* | 92 |
| 2.6 Animal diversity: vertebrates | 92 |
| 2.6.1 Amphibians | 94 |
| *Pitfall and fence trapping* | 94 |
| 2.6.2 Reptiles | 94 |
| *Pitfall and fence trapping* | 94 |
| 2.6.3 Birds | 94 |
| *Transect counts* | 94 |
| 2.6.4 Mammals | 97 |
| *Trapping for small mammals* | 97 |
| *Traps* | 97 |
| *Grid lay-out* | 98 |
| *Bait* | 98 |
| *Pre-baiting* | 98 |
| *Trapping* | 98 |
| *Handling of small mammals* | 98 |
| *Data handling and storage* | 99 |
| *Identification of vertebrates* | 99 |
| References | 101 |
| Appendix: Sample Arthropod Tally Sheet | 108 |
| Drafting team | 109 |
| Authors of sections | 109 |
| **Chapter 3: Freshwater Ecosystems** | 111 |
| 3.1 General Strategy | 111 |
| 3.1.1 Endangered freshwater ecosystems | 111 |
| 3.1.2 Biodiversity investigation in freshwater ecosystems | 112 |
| 3.2 Unique framework for the IBOY-DIWPA | 113 |
| 3.2.1 Freshwater environments | 113 |
| *Lakes* | 114 |
| *Rivers and streams* | 115 |

| | | |
|---|---|---|
| | 3.2.2 Target organisms | 115 |
| |     *Fish* | 115 |
| |     *Benthic organisms* | 116 |
| |     *Planktonic organisms* | 117 |
| 3.3 | Field and laboratory methods and data management | 117 |
| | 3.3.1 Bottom types and site selection for the littoral bottom survey | 117 |
| | 3.3.2 Water depth | 117 |
| | 3.3.3 Transect line | 118 |
| | 3.3.4 Physical structure and underwater landscape | 119 |
| | 3.3.5 Water column survey | 119 |
| | 3.3.6 Physico-chemical variables | 120 |
| | 3.3.7 Data from meteorological stations and remote sensing | 120 |
| | 3.3.8 Basic unit for field investigation | 121 |
| |     *Research vessel* | 121 |
| |     *Scuba divers* | 122 |
| | 3.3.9 Data management | 122 |
| 3.4. | Biodiversity survey for each organism | 124 |
| | 3.4.1 Fish | 124 |
| |     *Field Sampling* | 124 |
| |     *Field treatment* | 129 |
| |     *Laboratory procedures* | 132 |
| |     *Identification* | 133 |
| |     *Counts, measurement and ecological data* | 133 |
| | 3.4.2 Macrobenthos and macrophytes | 136 |
| |     *Field methods* | 136 |
| |     *Quadrat and mesh size* | 137 |
| |     *Underwater handling* | 139 |
| |     *Stone unit sampling for loose stones* | 142 |
| |     *Treatment of samples in the field* | 143 |
| |     *Laboratory methods* | 146 |
| | 3.4.3 Meiobenthos | 147 |
| |     *Field methods* | 147 |
| |     *Laboratory methods* | 148 |
| | 3.4.4 Epi-microphyte (attached algae) | 149 |
| |     *Field Methods* | 149 |
| |     *Laboratory methods* | 150 |
| | 3.4.5 Planktonic organisms | 150 |

| | |
|---|---|
| *Sampling from the water column* | 150 |
| *Collection and treatment of zooplankton* | 151 |
| *Collection and treatment of water sample, micro-, protozoo-, and phytoplankton* | 152 |
| *Treatment of bacteria* | 153 |
| *Treatment of algal picoplankton (APP)* | 154 |
| *Treatment of heterotrophic nanoflagellates (HNF)* | 155 |
| *Treatment of ciliates* | 156 |
| *Treatment of nano- and micro-phytoplankton* | 156 |
| 3.5 Examples of observation sites | 156 |
|   3.5.1 Clear lakes | 156 |
|   3.5.2 Turbid lakes | 157 |
| References | 158 |
| Drafting Team | 160 |

## Chapter 4: Latitudinal Biodiversity in Coastal Macrophyte Communities

| | |
|---|---|
| | 162 |
| 4.1 Introduction | 162 |
|   4.1.1 Importance of habitats | 162 |
|   4.1.2 Importance of monitoring | 163 |
|   4.1.3 Baseline studies | 164 |
| 4.2 Goals for Monitoring Coastal Ecosystems | 164 |
|   4.2.1 Latitudinal and related gradients | 165 |
|   4.2.2 Long-term monitoring | 165 |
| 4.3 Site Selection | 165 |
|   4.3.1 Regional level | 165 |
|   4.3.2 Local selection criteria | 166 |
|     *Infrastructure* | 167 |
|     *Baseline information* | 167 |
|     *Reasonably natural environment* | 167 |
|     *Long-term stability of the site* | 167 |
|     *Accessibility* | 168 |
|     *Biological character* | 168 |
|   4.3.3 Application of selection criteria | 168 |
|   4.3.4 Potential availability of no-fishing/no-take areas for stability of long-term monitoring | 168 |
|   4.3.5 Marine BioRap – Identifying biodiversity priority areas | 168 |
| 4.4 Sampling Protocol | 169 |

| | |
|---|---|
| 4.4.1 Sampling strategy | 169 |
| 4.4.2 Sampling methodology | 169 |
| *In-situ observation (non-destructive)* | 169 |
| *Direct removal (destructive)* | 170 |
| 4.4.3 Initial processing of direct samples | 170 |
| 4.4.4 Recommendations | 170 |
| 4.5 Subjects To be Studied and Monitored | 171 |
| 4.5.1 Species inventory of selected taxonomic groups | 171 |
| 4.5.2 Abiotic and biotic parameters | 172 |
| 4.5.3 Habitat and biodiversity mapping | 172 |
| 4.5.4 Species inventory and sampling | 173 |
| 4.5.5 All-biota taxonomic inventory | 173 |
| 4.6 Strategies for future activities | 174 |
| 4.6.1 Sampling kit | 174 |
| 4.6.2 Future activities | 174 |
| References | 175 |
| Appendix 1: Global 200 marine ecoregions occurring within the DIWPA region | 176 |
| Marine ecoregions: | 176 |
| *Large deltas, mangroves, and estuaries* | 176 |
| *Coral reef and associated marine ecosystems* | 176 |
| *Coastal marine ecosystems* | 177 |
| *Polar and subpolar marine ecosystems* | 177 |
| Appendix 2: Questionare form and invitational letter to be distributed at potential participating sites | 177 |
| DIWPA IBOY core or satellite site questionnaire form | 180 |
| Appendix 3: List of acronyms | 181 |
| Drafting Team | 182 |

## Chapter 5: Research Methods to Initiate PABITRA: The Island Ecosystem Branch of DIWPA

| | |
|---|---|
| Chapter 5: Research Methods to Initiate PABITRA: The Island Ecosystem Branch of DIWPA | 183 |
| 5.1 Introduction | 183 |
| 5.2 The DIWPA/PABITRA Relationship | 184 |
| 5.3 Underlying Theories for PABITRA | 186 |
| 5.4 Geoecology and Environmental Gradients | 190 |
| 5.5 Comparative Sampling of Upland Forests | 190 |
| 5.6 Sampling of Upland forests to Coastal Habitats | 194 |
| 5.7 Survey of Selected Taxa: The Hawai'i IBP Example | 195 |

5.8 Assessing Biodiversity Functions: Another Example from
the Hawai'i IBP ........................................................................... 199
   5.8.1 Sampling protocol of a large (80 ha) forest plot ........ 199
   5.8.2 Application of the guild concept ............................... 202
   5.8.3 Assessment of forest dynamics patterns .................... 202
5.9 PABITRA Scope for IBOY 2001 ............................................ 204
References ................................................................................... 206

# List of figures

1.1: IBOY candidate sites. 8
2.1: Pennsylvania style light trap modified for rainforest use. 46
2.2: The fully erected 'ground' Malaise trap. 48
2.3: Malaise trap 'tent' spread out to peg. 49
2.4: Malaise trap catch container 50
2.5: Frame for malaise trap lifted up to forest canopy. 50
2.6: Example of the catch from ground and canopy zones Malaise traps in tropical rainforest at Baitabag village, Madang, Papua New Guinea. 51
2.7: A window trap. 53
2.8: Example of the catch from window traps in ground and canopy zones in temperate deciduous forest at Tomakomai, Hokkaido, Japan. 54
2.9: A knockdown 'hoop' or collecting funnel. 55
2.10: Enlargement of sample jar holder in fogging hoop in Figure 2.9. 56
2.11: Sample results from a canopy knockdown sampling in tropical rainforest at Robson Creek, North Queensland. 57
2.12: Bark spray collecting hoop. 59
2.13: An ordinal profile from bark spray sampling of Knema sp. in mixed dipterocarp forest, Kuala Belalong, Brunei. 60
2.14: Pitfall trap. 61
2.15: Layout of pitfall traps; 1m between each. 61
2.16: An ordinal profile derived from pitfall trapping in subtropical rainforest in the Conondale Ranges, south-east Queensland. 62
2.17: Tullgren funnel array. 64
2.18: Sample results from a set of 10 litter samples from the Kau Wildlife Area, Madang, Papua New Guinea. 66
2.19: Abundances of soil mesofauna in a dry evergreen forest in Thailand. 68
2.20: Trap design (length unit = mm). 69
2.21: Vertical trap setting using a rope system. 70
2.22: Four collecting methods employed in 'Quadra-protocol'. 77
2.23: Collection of ants in the field. 79
2.24: Mounted ant specimen on a pin. 81
2.25: The principal microhabitats of termites in a forest. 82

# Figures

2.26: The standardized protocol for measuring termite species diversity, the belt transect method (from Jones and Eggleton 2000).   84

2.27: The standardized protocol for evaluating termite biomass, the pit-digging method.   86

2.28: Sample management and data analysis.   89

2.29: Pitfall trapping for amphibian and reptile survey.   96

3.1: Scuba diver with diving gear (after PADI Open Water Diver Manual, 1999)   123

3.2: Gill net (after Inoue, 1983).   125

3.3: Minnow trap. (after Yuma et al. 1997)   127

3.4: D-shaped hand net. (after Handa et al. 1987)   128

3.5: Color control patches.   132

3.7: Ekman-Birge sampler. (after Handa et al. 1987)   138

5.1: The island biogeography model of MacArthur and Wilson (1967).   186

5.2 Pacific area map outlining ten island regions (1–10) in a forest biogeographical order.   188

5.3: The PABITRA (Pacific-Asia Biodiversity Transect) Network as currently envisioned.   189

5.4: Comparative plot sizes used for vegetation studies in different landscapes.   192

5.5: Species/area curves for sites in Brunei. (Redrafted with slight modifications from Ashton 1965).   193

5.6: Idealized volcanic high island with near-natural landscape on left side and strongly human-modified landscape on right side.   196

5.7: Map of Hawai'i Volcanoes National Park showing design of the Mauna Loa transect sites 1-14 and location of the 80 ha forest dynamics plot as sampled by multi-disciplinary teams during the Hawai'i IBP.   198

5.8: Details of the 80 ha forest dynamics plot in the Kilauea rain forest.   200

5.9: Basic sampling unit of plant-biodiversity survey, a long, 6 x 100 m belt-transect, with subplots adapted in size for quantifying tree and undergrowth vegetation.   201

5.10: Guild spectrum of two arthropod communities in the Kilauea rain forest. For explanation of symbols see text. (After Gagne and Howarth 1981).   203

5.11: Vegetation map of the 80 ha IBP study plot in the Kilauea rain forest, derived from a color air photo taken July 1974. Original map scale before reduction was 1:1500. (After Cooray and Mueller-Dombois 1981). 205

# List of tables

| | | |
|---|---|---|
| 1.1: | The number of Protected Areas (PAs) in tropical regions. | 19 |
| 2.1: | Summary of required number of sites, required frequency and required period in the three types of protocol with different purposes. | 30 |
| 2.2: | Components of the census for IBOY in core and satellite sites. | 31 |
| 2.3: | Form for site descriptions and the census of environmental variables. | 37–8 |
| 2.4: | Form for the census of tall-trees (parameters of forest dynamics). | 40 |
| 2.5: | Form for the census of the other components (for core sites). | 41 |
| 2.6: | Form for the census of litter supply and decomposition rates (for core sites). | 43 |
| 2.7: | Form for the census of tree phenology (optional for core sites). | 44 |
| 2.8: | Species number collected by 4 methods. | 78 |
| 2.9: | Key works for termite identification in Asian region. | 87 |
| 2.10: | Sample date form for a regional list of vertebrates. | 95 |
| 2.11: | List of identification manuals for vertebrates. | 100–1 |
| 3.1: | Freshwater environments in East Pacific Asia and recommended methods for biodiversity investigations. | 114 |
| 3.2: | Sample field data sheet. | 130 |

# Preface

This book has been produced to assist in implementing the DIVERSITAS International Biodiversity Observation Year (IBOY) in the Western Pacific and Asian region. The purpose of IBOY in the Western Pacific and Asia is to observe the distribution of biodiversity and establish its status in the beginning of the 21st century. To achieve this, it was necessary to develop standardized methods for biodiversity assessment. These methods vary enormously between target organisms, ecosystems, and researchers. The first priority, therefore, for global monitoring of biodiversity is to standardize these methods in order to allow comparisons between sites. We greatly appreciate the efforts, cooperation and enthusiasm of the many scientists who have contributed to the preparation of this manual.

We recognise that this manual can still be improved and hope that many researchers will use and test it. We look forward to receiving constructive criticism about this edition; criticism aimed at producing improved editions of this manual in the future in the interests of improving the assessment of biodiversity for a wider range of organisms and ecosystems. It is our hope that this manual will promote monitoring of biodiversity and lead to a better understanding about the global status of biodiversity.

As explained in Chapter 1, IBOY was first proposed by the late Professer Tamiji Inoue as a cooperative activity among the researchers in DIVERSITAS Western Pacific and Asia (DIWPA). This has now been accepted as a core theme of the global DIVERSITAS program. We dedicate this book to the late Professors Tamiji Inoue, Takuya Abe and Masahiko Higashi of Kyoto University. Each was deeply committed to DIWPA and IBOY and, sadly, lost their lives in the pursuit of their research.

August 22, 2001<br>
Tohru Nakasizuka<br>
Nigel Stork

# Chapter 1: What is IBOY

*Chapter Editors: Nigel Stork & Tohru Nakashizuka*

## 1.1 Mission for IBOY-DIWPA

### 1.1.1 What is IBOY-DIWPA?

In 1997 the late Prof. Tamiji Inoue proposed the idea of a Biodiversity Observation Year (BOY) to mark the beginning of the 21$^{st}$ century and to increase research collaboration in the Western Pacific and Asian region. His proposal was to establish a latitudinal biodiversity inventory across the Asian region that would act as the baseline for future long-term monitoring of biodiversity. This idea received wide international support, both within the region and outside, and subsequently has been adopted and modified by the international coordinating committee of DIVERSITAS (see Box 1.1) as a global program: IBOY (International Biodiversity Observation Year). IBOY is a key activity of DIWPA: a year of observations on the status of the Earth's biological diversity at the beginning of the 21$^{st}$ century. So far more than 20 international projects have been nominated for IBOY. Inoue's original concept has been further developed as IBOY-DIWPA and is now one of the flagship projects of IBOY, entitled "Network study on biodiversity in ecosystems on the Green and Blue Belts in the Western Pacific and Asia." The aims, design and strategies for implementing IBOY-DIWPA are described below.

### 1.1.2 What are the goals of IBOY-DIWPA?

Although there have been many international calls to increase the study of biodiversity, inventorying, monitoring (see Box 1.2) and other basic research, very little of a practical nature has been achieved in the years since the signing of the Convention on Biological Diversity. To address this problem the IBOY-DIWPA program aims to:

1. Establish an international network of biodiversity study sites on latitudinal and other gradients in forests, freshwater and marine ecosystems in the Western Pacific and Asian region,
2. Provide an assessment of the state of biodiversity at these sites by inventorying, preserving, identifying and cataloguing biological specimens at these sites,

> **Box 1.1 DIVERSITAS and DIWPA**
> The DIVERSITAS Program was established as a joint initiative of the International Union of Biological Sciences (IUBS), UNESCO and the Man and the Biosphere Program in 1990 with a series of biodiversity focussed goals including the inventorying and monitoring of biodiversity and the ecosystem function of biodiversity. It aims to promote and integrate research on biodiversity resulting in syntheses for the scientific community and providing guidance for policy makers and resource managers concerned with the implementation of Conventions on Biological Diversity and Climate Change. DIVERSITAS is now supported by six international scientific organizations: IUBS (International Union of Biological Sciences), SCOPE (Scientific Committee on Problems of the Environment), UNESCO (the United Nations Educational, Scientific and Cultural Organization), IUMS (International Union of Microbiological Societies), ICSU (International Council for Science Union) and IGBP (International Geosphere-Biosphere Program). The organization is run by the Scientific Steering Committee which includes three representatives from the Western Pacific and Asia region. Detailed information can be obtained from: DIVERSITAS (*http://www.icsu.org/DIVERSITAS/Index/index.html*); IBOY (*http://www.nrel.colostate.edu/IBOY/*)
>
> In 1993 scientists in the Asian region created DIWPA (the International Network for DIVERSITAS in the Western Pacific and Asia) as a branch of DIVERSITAS to promote its goals in the region. DIWPA was established to (1) promote international collaborative research projects, (2) facilitate the international citizen program, and (3) promote governmental and non-governmental activities for the conservation and utilization of biodiversity. DIWPA aims to create an international network of the networks that already exist in each country and the research projects that focus on specific subjects or topics. More than 400 scientists from 41 countries and areas have joined DIWPA. DIWPA has already held a number of key regional biodiversity focussed international symposia and has contributed to greatly improved collaboration between biodiversity researchers in the region spanning from Russia to Australia. The six international training courses so far organised by DIWPA have played a particularly important role in improving understanding of biodiversity issues and developing collaboration. Detailed information can be obtained from: *http://ecology.kyoto-u.ac.jp/~gaku/diwpaindex.html*

3   Monitor the impact of climate change and other forms of environmental impacts on these ecosystems,
4   Undertake basic research aimed at understanding the relationship between biodiversity and ecosystem function,
5   Provide government and non-government organisations with information and tools to improve the sustainable management of biodiversity and biological resources.

A number of international meetings have been held with participants from many of the countries in the region to determine the goals of the IBOY-DIWPA program. It was agreed that i) the program will focus on the biodiversity of three major ecosystems in the Western Pacific and Asia: forests, freshwater and marine, and ii) that a manual or series of manuals would be prepared to provide standardized research methods; to aid the development of a biodiversity survey network in the region; and to ensure that the results of the biodiversity assessments are comparable across sites.

The strategy for IBOY-DIWPA is to develop methodologies to assess biodiversity at the site level along geographical and environmental gradients within the region, thus enhancing our understanding of the distribution of biodiversity and its role in ecosystem function. Data collection will open up new aspects of biodiversity for science and provide tools for ecosystem management. Below are some of the scientific questions that researchers will hope to answer:

> Why are forests green? Of about 1.6–1.8 million described species the dominant organisms are insects (ca. 750 000 species) and terrestrial plants (250 000). These species are particularly rich and abundant in forests. Perhaps up to 40 % of insects are phytophagous, so why are forests green when there are so many phytophagous species? In fact, forests are not always so green. Phytophagous insects in temperate forests sometimes consume significant amounts of leaves and even kill trees. Similarly, trees in tropical plantations sometimes suffer serious attack by insects.
>
> Why are there many more species in terrestrial ecosystems than in aquatic ecosystems even though the area of the land is much smaller than that of the oceans? Fungi are abundant in terrestrial ecosystems but scarce in aquatic ecosystems, especially in pelagic areas of oceans and lakes. Why is this the case?
>
> What are the linkages between biodiversity and ecosystem processes (or ecosystem function)? Does environmental change, such as deforestation, climate change, habitat fragmentation, lead to loss of

> **Box 1.2 International agreements on inventorying and monitoring**
>
> Recent international agreements and strategies have recognized the need to inventory and monitor biological diversity and have called on countries to initiate inventorying and monitoring programs:
>
> The Convention on Biological Diversity requires signatory parties to 'identify components of biodiversity important for conservation and sustainable use... and monitor, through sampling and other techniques, the components of biological diversity identified'. It also calls for signatories to 'identify processes and categories of activities which have or are likely to have significant adverse impacts on the conservation and sustainable use of biological diversity, and monitor their effects' and to 'maintain and organise... data derived from identification and monitoring activities'.
>
> Agenda 21 calls on signatories to 'undertake country studies or use other methods to identify components of biological diversity important for its conservation and the sustainable use of biological resources...' and to 'promote, where appropriate, the establishment and strengthening of national inventory... related to biological resources at the appropriate level'. Agenda 21 calls for the development of 'methodologies with a view to undertaking systematic sampling and evaluation on a national basis of the components of biological diversity identified by means of country studies' and to 'initiate or further develop methodologies and begin or continue work on surveys at the appropriate level on the status of ecosystems and establish baseline information on biological and genetic resources, including those in terrestrial, aquatic, coastal and marine ecosystems, as well as inventories undertaken with the participation of local and indigenous people and their communities'.
>
> The Global Biodiversity Strategy (WRI/IUCN/UNEP 1992) recommends that an early warning network be established to 'monitor potential threats to biodiversity and mobilize action against them' (Action 4).
>
> A global strategy addressing marine biodiversity (Norse 1993) suggests a global monitoring network on ecological processes to provide information for management and to provide warning of 'undesirable changes'.
>
> Inventory and monitoring efforts are also central to the activities described in Guidelines for Country Studies on Biological Diversity (UNEP 1993). Country Studies provide an information base from which nations can develop biodiversity action plans, as requested by the Biodiversity Convention.

biodiversity and ecosystem functions such as herbivory, pollination, seed dispersal, and nutrient cycling?

Are some groups of organisms more affected by disturbance and change than others? What are the attributes of these organisms that make them more vulnerable?

Can we find suitable sets of organisms that can be more readily sampled and understood than others and which will act as indicators or 'predictor' sets (Kitching 1993, Kitching et.al. 2000) for biodiversity and environmental change?

### 1.1.3 Strategy for implementing IBOY-DIWPA

This manual was discussed at a meeting in Kyoto 1998, and editing commenced soon thereafter. The first version was completed in July 1999 and reviewed by the DIWPA committee. Two parts of this Manual in particular now have been revised ahead of the other parts: first, the Mission for IBOY-DIWPA, and second, a Manual for Biodiversity Assessment in Forests. In the year 2000 and 2001, two training courses were carried out as pilot studies for forest assessment using the forest manual. The usefulness and applicability of the Manual was tested in these training courses.

Not all participating scientists and countries are at the same level of preparation for IBOY-DIWPA and it is expected that not all site assessments should commence from the beginning. Following observations throughout the region in 2001 and later years, we will move on to the analyses of the results obtained. This will take several years since the biodiversity of the region is very considerable. IBOY-DIWPA will hold a Biodiversity Summit in the Western Pacific and Asia in 2005 to discuss the conclusions of the program. This meeting will include the public, policy makers, and scientists. We believe our results will contribute greatly to the international discussion on biodiversity and its conservation. We expect the activities of IBOY-DIWPA will provide the baseline and start of long-term monitoring of biodiversity in the region (see Box 1.3).

### 1.1.4 Integrated approaches to biodiversity assessment and capacity building

IBOY-DIWPA aims to improve the effectiveness and efficiency of national and international biodiversity assessment effort (see Stork & Sherman 1995) through collaborative, multidisciplinary and multisectoral research. Co-ordination through IBOY-DIWPA is essential if resources, including time, money and personnel are to be used to maximal effect. Good co-ordination prevents certain geographical areas from being neglected, while also avoiding

> **Box 1.3 IBOY-DIWPA Timetable**
>
> *1998–2000*
> Development of IBOY-DIWPA program concept
> Increasing participation from countries and scientists within the region
> Study training program
> Seek funding for training, program coordination
> Pilot studies
> Development of protocol manuals and testing
> Selection of field study sites
>
> *2000–2003*
> Commencement of biodiversity assessment of field sites
> Development of network and data management
>
> *2001–2005*
> Analyses of samples
> Data analysis and hypothesis testing
>
> *2005–2006*
> Biodiversity Summit in Western Pacific and Asia
> Other International meetings
>
> *2006*
> Implementation of policies and manuals

repetitive efforts being made in others. Much of the scientific data on biodiversity is difficult to synthesize because of its fragmentary character, limited comparability and lack of ready accessibility.

One of the problems that IBOY-DIWPA faces is the lack of good taxonomic collections in the region and the staff required to undertake systematic studies. If inventorying and monitoring of biodiversity in the region is to be increased to the levels suggested by international agreements such as the Convention on Biological Diversity and Agenda 21 or by scientific programs such as DIVERSITAS (IUBS/UNESCO) or Systematics Agenda 2000, a vast increase in systematic and ecological infrastructure and human resources in the region will be required. However, such capacity and infrastructure exceeds the capabilities of even the most advanced nations, and biodiversity is not restricted by political boundaries. There is therefore a need for collaboration, to share expertise and resources.

One way IBOY-DIWPA will improve the skills base in the region is by offering specialist training courses in, for example, biodiversity assessment of forest invertebrates or vertebrates, data management and sharing, statistical procedures for biodiversity analyses, collection management, and the systematics of different groups of organisms. DIWPA has held several such courses, with students and lecturers from many countries participating, thus increasing collaboration, cooperation and understanding between scientists. IBOY-DIWPA will continue to develop this program of training courses.

Few countries have the tools required to identify their fauna and flora or clear programs for inventorying these organisms. If museums, herbaria and other institutions charged with the task of providing systematic services for inventorying are to be able to carry out these tasks, they need better resources, their collections need to be better maintained, and a new generation of professional researchers and technicians trained and funded (see Stork 1995). The same is also true for monitoring biodiversity. In many countries the monitoring of biodiversity is almost non-existent.

### 1.1.5 Study sites

The costs of biodiversity assessment can be very considerable. It is therefore essential that careful consideration be given to the selection and number of sites for such assessments. Decisions on these issues are, of course, matters for individual nations and scientists. Here we offer guidance for the important selection criteria that will facilitate and aid the goals of the IBOY-DIWPA program. Many countries are considering how to implement biodiversity assessments as part of national biodiversity strategies and how to research the impacts of climate change and other forms of environmental change. Collaboration through IBOY-DIWPA provides an unparalleled opportunity to examine and understand regional patterns of biodiversity and the impact of such changes. However this will only result if sites are selected in a similar way and if standardised methods of biodiversity assessment are used as recommended in the following four manuals for forest, freshwater, marine and island ecosystems. Details on site selection (see Fig. 1.1) will be discussed in the chapter on each ecosystem.

### 1.1.6 Selection of target organisms

It is impractical to assess all of the biological elements of ecosystems, including bacteria, protists, fungi, plants and animals, with equal thoroughness, simply

**Biodiversity Research Methods**

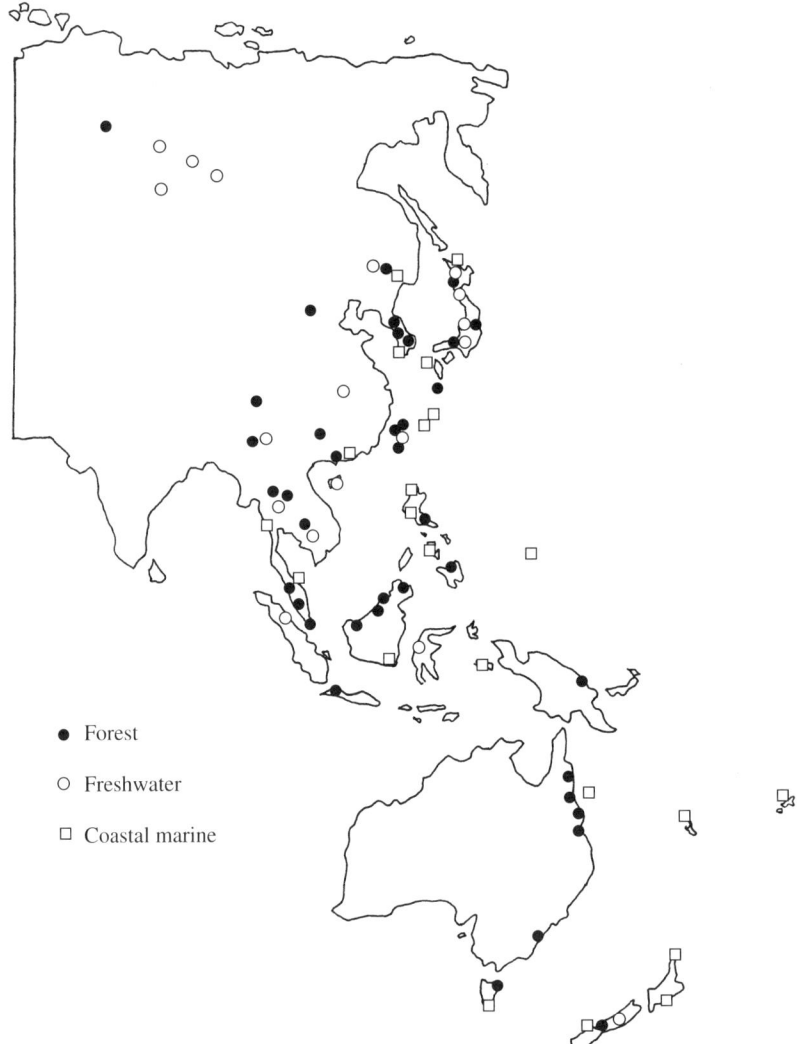

*Figure 1.1: IBOY candidate sites.*

because there is not enough money and scientific expertise to do this. There has been much discussion about what criteria to use for selecting indicator taxa and on the usefulness of indicator taxa (eg. Lawton et. al. 1998). Pearson & Cassola (1992), for example, proposed seven criteria for selecting indicator taxa for the quantitative assessment of biodiversity; 1) well known and stable taxonomy, 2) well known natural history, 3) readily surveyed and manipulated, 4) higher taxa broadly distributed geographically and over a breadth of habitat types, 5) lower taxa specialized and sensitive to habitat changes, 6) patterns of biodiversity reflected in other related and unrelated taxa, and 7) potential economic importance.

Which groups of organisms are important differs depending on the ecosystem. For example, forests are characterized by the super dominance of plants, especially trees. Many terrestrial plants depend on animals for pollination and seed dispersal, while algae in aquatic ecosystems depend on the water body for the dispersal of propagules. Thus we mainly deal with plants and animals in above ground subsystems of forest ecosystems. The situation is quite different in pelagic areas of aquatic ecosystems. Here the microbial loop plays an important role in the ecosystem process. It is therefore important for researchers to identify key groups of organisms for particular ecosystems, or predictor sets of indicator groups.

## 1.2 Mission for biodiversity assessment

### 1.2.1 Why is biodiversity assessment essential?
The goals of IBOY-DIWPA focus on the assessment or inventorying and monitoring of biodiversity at a network of sites throughout the Western Pacific and Asian region in participating countries. They also require the use where possible of standardised methods of assessment in order to allow comparisons between sites and to provide a broader regional assessment of biodiversity. Section 7 of the UNEP Global Biodiversity Assessment project (UNEP 1995) provided for the first time a real discussion of the scientific and practical issues concerning inventorying and monitoring of biodiversity (Stork & Samways 1995a). The following text is a condensed version.

Biodiversity assessment can comprise either or both of two elements: inventorying and monitoring.

*Inventorying* is the surveying, sorting, cataloguing, quantifying and mapping of entities such as genes, individuals, populations, species, habitats, biotopes, ecosystems and landscapes or their components, and the synthesis of the resulting information for the analysis of processes. Inventories provide snapshots of the state of biodiversity at particular locations. Inventories also provide baseline information for the assessment of change and can be used to identify key variables and bioindicators for monitoring.

*Monitoring* is intermittent (regular or irregular) surveillance to determine the extent of compliance with a predetermined standard or degree of deviation from an expected norm. Monitoring is usually goal oriented, telling us how things are changing, and is repeated at intervals. Monitoring of biodiversity partly aims to develop a strategic framework for predicting the behaviour of key variables to improve management, increase management options and to provide an early warning of system change.

The success of monitoring depends on a number of factors including
- Being clear on the spatial and temporal scales of investigation and management;
- Having a keen awareness of the time scale of a project and its feasibility within that time frame;
- Using an appropriate taxon or various taxa to provide the information needed to illustrate pertinent changes;
- Using methodologies, including statistics that are appropriate and efficient to the study or management site, while also providing comparable results from other sites locally, regionally or globally;
- Standardizing data collections and statistical analyses;
- Maintaining voucher collections of the subject organisms;
- Using existing data where they are valid; and
- Assuring not only the appropriate biotic, but also abiotic and human variables are recorded, and put into the appropriate statistical framework.

It is important to recognise the linkages between inventorying and monitoring and hence the linkages between basic science and management. Inventories provide the base line while monitoring measures *change* and guides management. Inventorying and monitoring of biodiversity provide fundamental and essential biological information used by many basic and applied scientific disciplines including systematics, population biology, behaviour, ecology, biotechnology, soil science, agriculture, forestry and fisheries sciences, conservation and environmental sciences (see also Stork & Samways 1995b). They provide information which may be used to provide a basis for the scientific research which is necessary to understand the world in which we live, to define the current and future options available for meeting human needs, and to guide immediate and long-term management, policy and decision making (see Box 1.4).

Scientists know enough about overall distribution of threats to biodiversity to be able to guide the allocation of resources to support biodiversity assessments and recommend direction for sustainable resource use. Experience shows that where resources are wisely and systematically used to inventory and monitor biodiversity, challenges can be met. An example is the application of adaptive management strategies, which hold promise for achieving sustainable use of fishery resources and the maintenance of biological diversity. Risk assessments

> **Box 1.4 Areas in which inventories and monitoring activities are important include**
> - Providing information for determining and conserving biological diversity;
> - Providing information necessary for the sustainable management of natural resources;
> - Identifying economically valuable products from wild species (bioprospecting);
> - Maintaining or increasing the productivity of agricultural systems through the identification of (i) new varieties or new species of use to humans and (ii) beneficial and harmful organisms;
> - Improving human health through the identification of pest organisms and beneficial organisms;
> - Understanding ecosystem processes so that the ecological services essential for human survival can be maintained;
> - Defining the impact of human activities on biodiversity so as to help reduce undesirable effects on the environment;
> - Understanding the potential effects and impact of climate change and other forms of natural environmental change;
> - determining the aesthetic benefits of diversity so as to preserve the quality of human life.

that consider availability of resource abundance and productivity have become an integral part of scientific advice to managers. Among the relatively new fisheries for which access and fishing effort have been controlled from the start of fishing, leading to sustainable biomass yields, are those of the Falklands Islands and bottom fisheries of the north-west shelf system of Australia.

The identification and monitoring of biological diversity are regarded as being essential for supporting articles 8, 9 and 10 of the Biodiversity Convention. Glowka *et al.* (1994) point out that the results of identification and monitoring projects are important for:
- Developing strategies, plans and programs (article 6(a));
- Integrating conservation and sustainable use into sectorial and cross-sectorial plans, programs or policies (article 6(b));
- Undertaking environmental impact assessment (article 14(a) and (b)); and
- Negotiating access agreements, including benefit-sharing (article 15(7)).

Actions affecting biodiversity and natural resources are often based on inadequate information as the data necessary for informed decision-making often are unavailable, incomplete or unreliable, or not presented in a format which policy-makers and managers can use, and frequently the data are incorrectly interpreted. Given the limited funds for data collection, analysis of *which* information is useful to decision-makers (defining the audience and determining their needs) is critical for knowing which taxa are to be inventoried and monitored. These issues need to be carefully considered when deciding, for example, which sites and which taxa should be studied, at what scale and how frequently. Some of these issues are discussed further below.

## 1.2.2 How frequently and where should biodiversity assessments be made?

The frequency and geographical scales for biodiversity assessment and monitoring depend on the particular problems to be addressed. For example, two major areas where monitoring is applicable across varying geographical scales are:
- The assessment of the effectiveness of biodiversity management. For national, regional or park programs designed to preserve or increase biodiversity or to optimize biodiversity in systems managed for other goals (e.g., plantations), monitoring is a way of documenting whether the stated goals were achieved;
- The need for an early-warning system of impending adverse change. The objective here is to develop a monitoring system that is sensitive to unusual changes in biodiversity *before* they become too critical.

Similarly, the kinds of problems that are being examined, the level of assessment, and the groups of organisms or perturbations to ecosystems will determine the timescale for inventorying and monitoring. For some questions it may be necessary to carry out inventories on a daily basis to monitor daily changes, whereas for others it may be sufficient to carry out inventories every year, ten years, or longer.

## 1.2.3 At what biodiversity level should assessments be made?

The answer to this question again depends upon the problems and scientific questions being considered and the possible applications of the information to be gathered. Examples of how such assessments provide important and useful information at each level of biodiversity are suggested below.

### At the genetic level

Biochemical and molecular data are becoming increasingly useful for assessing and managing regional diversity in natural populations. A conceptual framework for developing conservation strategies is emerging from the recently-developed field of 'conservation genetics'. Numerous genetic techniques are now available for characterising wild and cultivated populations or populations managed in captivity (Hillis & Moritz 1990; Avise 1994). Molecular methods can be used to survey genetic variability at three levels of organization: individual organisms, comparisons within a particular species, and comparisons among different species.

At the first level, genetic surveys can identify individuals that may have suffered reductions in genetic variability because of large reductions in population size. Reduced levels of genetic variability can often lead to inbreeding depression, especially in species whose life-history patterns have historically produced large population sizes. Inbreeding depression may be due to the homozygosity of recessive deleterious genes or to a reduction in genetic variability itself, and may be important in small populations.

Geographic surveys are important for identifying appropriate groups of genetically-similar populations for conservation or management. In migratory species, such as sea turtles and salmon which home to natal areas to breed, genetic methods can be used to identify the origins of individuals from threatened breeding areas that are harvested or caught incidentally in high-seas drift-net fisheries. In devising a conservation strategy for a threatened species, it is important to conserve not just a few representatives of a species, but also to preserve a wide range of the species' genetic variability. Identification of genetically distinct groups is especially important in the *ex-situ* management of captive populations in zoological gardens or of artificially propagated semi-wild populations.

Genetic methods are also important for addressing problems at the species level. Genetic data can be important for identifying morphologically-cryptic species which may go unrecognized or which may be inappropriately mixed with genetically unrelated populations. The systematics of many commercially-important plants and animals are still poorly understood. Such information can be used, for example, to search for genes in closely-related species that may be used to increase production or to enhance disease resistance. Phylogenetic studies of cultivated species and their wild relatives are important for identifying the closest relatives of domesticated strains to preserve important genetic resources. The effective enforcement of international agreements on the trade of rare and threatened species (such as CITES) will also require better information on species limits and identifications (Pain 1994). Genetic data may also be of value in policing the purported 'captive breeding' of species of high economic value.

Genetic data are also important for genetic engineering. For example, by synthesizing naturally-occurring polyploid species it can be shown how such polyploids may have evolved from their putative diploid ancestors.

## At the population level

Many of the above genetic arguments relate also to inventorying and monitoring at the population level. The monitoring of population dynamics can result in the development of sound conservation strategies for particular systems and produces generalizations of potentially wide applicability (Soulé & Kohm 1989). Such monitoring must be long-term if it is to generate data that will enable detection of a signal (such as declining population size) over the noise (or random fluctuations) inherent in most systems. Conservation recommendations based on short-term studies often have to be reconsidered in the light of subsequent information. For example, rare events such as a drought or an epidemic may be the key to the long-term population dynamics of a species (Soulé & Kohm 1989).

The identification and monitoring of populations provide important information for the characterization of rarity and threat status. The new IUCN Red List categories (IUCN 1994) include population status and rate of decline among the criteria for categorizing species as 'Critically Endangered', 'Endangered' or 'Vulnerable'.

Demographic information is also important in determining Minimum Viable Population (MVP) and Population Viability Analysis (PVA), two approaches for dealing with the problems of small population size. The MVP is the population size which provides a given probability of persistence of the population for a given amount of time (e.g., a 95 % expectation of persistence without loss of fitness for several centuries), while the PVA provides the probability that a population will persist for some designated period.

Population inventorying and monitoring provide information also for determining sustainable harvest and yields. Harvesting is essentially a consumer-prey relationship where humans are the consumers. Humans harvest a vast range of life forms (e.g., molluscs, fish, trees, game animals, herbs) and, in most cases, exploitation involves the complete removal of the individual. Less commonly, parts or products of the organisms are harvested (e.g., leaves, fruits, resins, wild honey, bird feathers). Analyses of harvest and yield involving individuals may be quantitatively similar to those involving organism parts or products, because partial harvest may often affect individual survival and demographic traits. Models of sustainable harvesting require both time series of population abundance (relative or absolute) and estimates of demographic parameters (Clark 1981; Walters 1986; Getz & Haight 1989; Royama 1992). Substantial yields are most difficult to sustain if a population is fluctuating. It is important also that models are not based on unrealistic assumptions, such as a constant environment.

## At the species level
Scientific information resulting from species inventories supports a wide range of activities. For example:
- For a nation, knowing the identity and geographical distribution of its species is perhaps the most important information available in its attempts to preserve and use its biodiversity. This knowledge comes primarily from inventories of various kinds and provides basic knowledge for day-to-day management decisions regarding natural resources.
- Species inventories across global, regional or national levels are important for establishing patterns of diversity and identifying areas of endemism, two primary components for evaluating areas for protected status.
- At more local levels, species inventories are required for meeting management objectives for protected areas (UNEP 1993), including maintaining ecosystems and biological diversity, conserving genetic resources, monitoring change, providing for sustainable economic uses, such as hunting, fishing, recreation, tourism, and forestry, and facilitating education and research.
- New taxa and new biological material of described taxa collected during the inventory process are essential for clarifying species identifications and relationships (Mound and Gaston 1993). Inventories of small invertebrates or micro-organisms often contain a high proportion of undescribed species.
- Inventory activities result in the accumulation of scientific specimens, tissues and other biological material, preserved or living, which help to build the natural history collections of museums, botanical gardens and zoos and provide a basis for scientific studies and conservation management well into the future. Tissues and other living material increase the holdings of gene and tissue banks, seed banks, culture collections and other institutions concerned with genetic resources.
- Studies of specimens collected during inventories yield data useful for resolving the phylogenetic relationships of species, which, in turn, are essential for building predictive classification systems and permit the estimation of character diversity for comparison of biota (Williams *et al.*

1995). These relationships can be used to help prioritise areas for conservation or other land management decisions (e.g., Vane-Wright 1993).
- Scientific reports, papers, distributional lists and systematic monographs that are produced following systematic analysis of natural history collections, may, in turn, form the basis for many products useful to a general audience, including pamphlets, field guides and photographic guides.
- Information from species inventories may be used in biogeographical analyses to understand patterns of diversity, the processes that have produced the Earth's biological variety and present-day distributions and to predict how biodiversity may change with altered environments. Biogeographical information also helps solve biological challenges such as discovering the origins of agricultural pests and diseases, leading to the identification of their natural predators and parasitoids.
- Inventories provide ecologists with data about the presence or absence of species in an area, which is essential for understanding community structure, function and processes.
- Inventories and the collections they produce also expand the scientific study of the region in question. Information from specimens and from field notes can be incorporated into Geographic Information Systems (GIS) to provide a more sophisticated basis for resource management. Furthermore, information associated with specimens, including behavioural and ecological data, are an essential record available to all scientists that were not part of the inventory process itself.
- Species inventories provide the foundation for future industrial applications, particularly those associated with bioprospecting (Reid *et al.* 1993). Likewise, the discoveries that result from inventories can support agricultural and forestry programs.
- Finally, inventories contribute to the education and training of scientists and, therefore, help build biodiversity science capacity. Moreover, in the long run, inventories invariably contribute to general public education about biodiversity and conservation.

## Species turnover in ecosystems

Changes in species diversity occur naturally over time in all communities and ecosystems. Human disturbance may change the rate and/or direction of these changes. However, because the dominant organisms in many ecological systems are long-lived, many important changes in communities and ecosystems are too slow for us to sense directly. Thus only long-term monitoring may reveal such important changes (Hellawell 1991; NRC 1993). Species-based monitoring programs can document natural patterns of change (positive or negative), or lack of change, to establish a baseline for understanding the impact of natural disturbance on species composition and abundance in ecosystems. Once a baseline has been established, it can be used to detect changes in biodiversity (variations from the norm) that result either directly or indirectly from human disturbance. Monitoring can also be used in a predictive manner to test anticipated change, for example as a function of global warming models (Solbrig 1991). Thus, by knowing how different species respond to different stresses, we can predict the extent of local species extinction and replacement and, therefore, ecosystem change (Soulé and Kohm 1989). It is also important that we identify such changes early enough that corrective action can be taken while multiple options are still available. Reductions in the number of options increase the expense of remedial action.

## At the ecosystem and landscape levels
Inventorying ecosystems and monitoring changes in their character and processes are essential for distinguishing short-term variation from long-term patterns (Chernoff 1986). Ecosystem monitoring is important for management and policy making in many ways, including:
- Understanding ecosystem processes;
- Understanding the effects of landscape fragmentation, habitat destruction, and other forms of disturbance;
- Distinguishing where population trends are due to natural fluctuations as opposed to anthropogenic factors;
- Predicting the possible effects of global climate change;
- Predicting the ecological changes that may follow the extinction of one or more key species or taxonomic groups;
- Measuring changes in earth cover and land use and the impacts of these on biodiversity.

Land usage can change rapidly, particularly in agricultural and urban landscapes (Forman and Gordon 1986). Also, ecosystem limits can extend or shrink over time even within protected areas. The problem of 'shifting mosaics', especially within protected areas, can be better understood with information on the distribution of different biotope and ecosystem types from remote sensing. The dynamic characteristics of a watershed, for example, are best inferred through this process and the resultant maps provide information on features such as changes in levels of water bodies and river courses. Likewise, the extent of littoral vegetation on beaches, soil binders in sand dunes and mangrove vegetation along estuarine shorelines, all of which are important in regulating soil erosion, can be represented and monitored with remote sensing.

Within the nearshore areas and extending around the margins of global land masses, species in coastal ecosystems are being subjected to increased stress from toxic effluents, habitat degradation, excessive nutrient loadings, harmful algal blooms, emergent disease fallout from aerosol contaminants, and episodic losses of marine resources from pollution effects and overexploitation. The long-term sustainability of biodiversity of resource species in coastal ecosystems as sources for healthy economies appears to be diminishing. A growing awareness of problems such as these has accelerated efforts to assess, monitor, and mitigate the stresses on coastal ecosystems (Sherman 1994).

## 1.2.4 The rationale for inventorying and monitoring inside and outside protected areas

Monitoring in protected areas aims to provide managers with the information they require for effective management and to provide information as part of national and international monitoring programs. This means that each protected area must have a program to identify the monitoring needs and available resources and to specify the type of monitoring activity, including where and when it should be carried out. It also necessitates guidance and support from the appropriate national and international authorities, specifying which programs to follow and identifying appropriate methodologies (Lucas 1992). Although these stipulations were originally formulated for protected areas, they are equally applicable to many areas without such protection (Hale & Lamb 1997).

Protected areas have a legal designation and, usually, a set of management objectives set out either in legislation or in a management plan. Programs are needed to monitor the degree of implementation of these objectives and the effectiveness of management practices in order for action to be taken to correct both inappropriate management and actions that are having unforeseen results. It may be sensible also to have the involvement of independent advisors who assist monitoring activity or carry out periodic 'peer' review. Monitoring in protected areas assists the development of management strategies (Spellerberg 1991). For example, data from aerial photographs and visitor surveys were used to quantify changes and provide a database for long-term monitoring of Rondane National Park in Norway (Fry *et al.* 1992).

It is also necessary to develop and adopt internationally accepted guidelines and criteria for assessing management effectiveness. Once developed, the agreed methods/guidelines need to be tested further, disseminated, and their implementation encouraged.

## Status of identified features in protected areas

Most protected areas are established to conserve key features, whether these are species, biotopes, ecosystems or landscapes. There is a need to monitor the status of these features using appropriate methods in order to manage them effectively. The development of improved monitoring programs in protected areas is important. However, not only do many protected areas lack monitoring programs, many do not have even basic species inventories. In a study of literature relating to 8,715 protected areas in tropical forest countries, WCMC was able to locate inventories of major taxonomic groups for only 5 % of the sites (Table 1.1).

## Threats to protected areas

Many protected areas are under threat. The majority of 135 national parks in more than 50 countries surveyed in 1982 reported threats of one type or another (Machilis & Tichnell 1985). More recent surveys show no improvement in the situation and,

Table 1.1: The number of Protected Areas (PAs) in tropical regions, the percentage with species inventories (I) (based on limited field observations, partial surveys of the area and systematic surveys), and the percentage with comprehensive species inventories (C) (based on systematic surveys only) shown in brackets.

### A) PLANTS

| Region | No. PAs[1] | Higher Plants I | Higher Plants C | Trees I | Trees C |
|---|---|---|---|---|---|
| Africa | 2,543 | 1.6 | (0.6) | 1.2 | (0.4) |
| Asia | 3,599 | 2.2 | (0.4) | 1.3 | (0.3) |
| Latin America[2] | 2,413 | 1.1 | (0.3) | 0.3 | (0.0) |
| Pacific | 160 | 9.4 | (7.5) | 1.3 | (0.0) |
| Combined | 8,715 | 1.8 | (0.6) | 1.0 | (0.2) |

### B) ANIMALS

| Region | No. PAs[1] | Mammals I C | Birds I C | Reptiles I C | Amphibian I C | Fish I C | Butterflies I C |
|---|---|---|---|---|---|---|---|
| Africa | 2,543 | 4.3 (0.3) | 3.8 (0.9) | 1.2 (0.2) | 0.7 (0.0) | 0.1 (0.0) | 0.1 (0.0) |
| Asia | 3,599 | 3.3 (0.3) | 4.8 (0.8) | 1.0 (0.1) | 0.7 (0.1) | 0.1 (0.0) | 0.0 (0.0) |
| Latin America[2] | 2,413 | 1.4 (0.1) | 1.7 (0.3) | 1.1 (0.1) | 0.7 (0.1) | 0.0 (0.0) | 0.0 (0.0) |
| Pacific | 160 | 6.3 (5.6) | 9.4 (9.4) | 5.0 (4.4) | 0.0 (0.0) | 0.0 (0.0) | 0.0 (0.0) |
| Combined | 8,715 | 3.1 (0.3) | 3.7 (0.9) | 1.1 (0.2) | 0.7 (0.1) | 0.0 (0.0) | 0.0 (0.0) |

1 All categories of protected area
2 Including Mexico, Central and South America and Caribbean

(From: Murray et al. 1992, see also Stohlgren et al. 1995).

therefore, it is important that such threats to protected areas and the effects of those threats be identified, monitored and, hopefully, eliminated or mitigated.

## Use and socio-economic benefits of protected areas

Protected areas provide a range of services and benefits to people. Important information for managing these areas is provided by monitoring programs which address what resources are used, either consumptively or non-consumptively, and what services are provided. Variables can then be identified and used to monitor both the health of the resource and its value (usually economic or social). One of the easier value variables to monitor is visitation, but it may be necessary to monitor a much wider range of variables to more completely assess the value of protected areas to society.

For example, a report has recently been prepared for the Great Barrier Reef Marine Park Authority, which assessed the economic and financial values of the Great Barrier Reef during 1991–2 (Driml 1994). Crudely summarized, tourism is estimated to be worth A$ 682 million, commercial fishing A$ 128 million, recreational fishing and boating A$ 94 million and research A$ 19 million, for a total of A$ 923 million. This research was extended to cover a number of other areas, with the total value of tourism and recreation in six major protected area complexes estimated at over A$ 1.9 billion. This is clearly a substantial input to the economy of the regions concerned, underlining the importance of understanding the value of these areas. A range of intangible benefits of course, also augments this value, such as maintaining local to global processes intact.

## Inventorying and monitoring outside protected areas

Natural areas not designated as protected face the pressure of being converted to other land uses such as cash-crop production (Hale & Lamb 1997). Major portions of a nation's forests are, for example, generally earmarked as timber-production areas (Harris 1984). Sound management of these areas requires that an inventory be carried out to ascertain what resources are available and in what quantities, and the inventory repeated at regular intervals – for example, 10 years – to provide information on sustainable management of these resources. Inventorying and monitoring in such national programs are almost always targeted at timber species, and, in some cases, at non-timber species such as rattan.

The global level of primary productivity to support the present annual yield of fishery resources has been reached (94 million metric tons). It is likely that substantial unmanaged increases in fishery biomass yields will be obtained by fishing down the foodweb from fishes, causing losses in biodiversity at the species abundance level (Pauly & Christensen 1995). The need for monitoring these changes at the global level in relation to cascade effects through the foodweb of marine ecosystems is recognized as an important contribution to the sustainable management of fisheries (Beddington 1995).

## 1.2.5 Scales, planning and approaches to inventorying and monitoring

Assessments of the changing demographics of biodiversity around the globe require comparable and quantifiable biological and environmental data related to the causes and effects of the changes. And yet, the time, effort and funds required to inventory and monitor global biodiversity will need to be based on the most cost-effective and efficient scientific methodologies presently available.

A more detailed account of the methods for compiling inventories and for monitoring different levels of biological diversity are presented in Samways *et al.* (1995). This paper also considers the taxa and ecological indicators that might be used as foci for research and management and provides detailed discussion on points and areas that are making particularly important contributions to biodiversity assessment using new technologies.

The approaches and methods adopted for inventorying and monitoring depend on the level of diversity being assessed and the geographical scale of the analysis. It is therefore essential that the goals of a project are clear and relate to the problem being considered. The value of biodiversity findings depends on the relevance and importance of the goals set and the thoroughness of the execution of the project, as well as its accuracy. The feasibility and accuracy of inventorying and monitoring programs also depend on the availability of adequate resources. The accurate interpretation of biodiversity information also requires an understanding of the ecological dynamics of the geographical area.

Targets for inventorying and monitoring include genes, populations, species, communities, biotopes and ecosystems. Variables used to monitor species can be compositional (abundances, cover values, densities, biomass, etc.), structural (dispersion, range, population structure, biotope variables, etc.), or functional (demographic processes, metapopulation dynamics, population genetics, growth rates, etc.).

Historical collections of organisms provide extremely important baseline information on how a species has changed its range, abundance and form over time. Other existing data may also provide useful historical or even spatially comparative information.

Species inventories must use standardized methods wherever possible and use recognized sampling protocols that sample the appropriate species adequately but not excessively. This makes the data and the analysis comparable with other studies in other areas. Standardization also allows for later validation and calibration of findings from one time or place to the next. In this respect, sampling protocols and standards such as those produced by Heyer *et al.* (1994) are needed for target taxa. A statistical framework for the analysis of biodiversity information is also essential and should be planned prior to data collection. This then provides meaningful and efficient fact gathering and interpretation.

Voucher collections are essential for verification of field data and provide a permanent historical record. In this sense natural history collections are a vital source of baseline data for inventorying and monitoring.

The methods employed in inventorying and monitoring vary according to the level of biodiversity and the geographical scale of the study and may include checklists of species, relative abundances, population densities or complete counts of individuals. Methods should also take into account behavioural, developmental, and seasonal variation in abundances. Techniques for monitoring species are highly diverse, and the chosen method is often a compromise between the ideal and what can actually be achieved. Censuses are important for some large and conspicuous organisms, and mark-recapture techniques may be an important means of estimating population size. Similarly, territory mapping, point counts, and transects can be valuable, particularly over a long time period. Red Data Books and Atlases also provide valuable long-term information on changes in rarity status.

Population surveys either focus on population size or on demographic aspects of population structure, the latter being particularly important in understanding how a population is changing in the face of anthropogenic disturbance. There are many methods for estimating population size, chosen according to the organism and level of accuracy required, although none of the methods is entirely satisfactory. Nevertheless, in the case of large mammals fairly accurate estimates of population size over time are often essential for their optimal management. Furthermore, for all organisms an estimation of population size is important for determining rarity or threat status. For very small populations, it is essential that their minimal viable population (MVP) be estimated as well as the chances of the population persisting, using population viability analysis (PVA). An important aspect of population monitoring relates to harvestable organisms. Monitoring and modelling determines optimal and sustainable utilization.

Remote sensing systems have become an important means of monitoring key aspects of vegetation diversity, both relative to wildlife and to human populations. Each type of imagery however, has its own uses, advantages, and disadvantages. The large-scale distribution of vegetation can be determined from satellite imagery. Seasonal and multispectral imagery is useful for determining composition and condition of the overstorey. Structure and biomass usually require some height estimates, such as with stereo imagery, radar, or laser profilers. Rapid updates of conditions of small areas can be done with airborne videography or digital cameras. Global positioning units are useful in linking remote sensing and field plots with a Geographic Information System. Remote sensing's quick areal coverage at relatively low cost may, however, require supplementary information obtained on the ground.

On a larger scale, the marine environment is internally moving and distinctly three-dimensional; it therefore presents special monitoring problems. Also on the larger scale, monitoring is used in inventories on biotopes, ecosystems and landscapes. Such ecological units are sometimes designated a name on the basis of their dominant species. Monitoring of species plays a major role in assessing changes to ecological systems over time. This has management value and is important in biodiversity protection. At the grossest level, biodiversity inventories will target landscapes, which are the composites of alpha, beta and gamma diversities. Similarly, monitoring biodiversity and its components provides information on land use dynamics and on how to manage large marine ecosystems and freshwater systems. Also important is the inventorying of biota in protected areas, and the monitoring of its status over extended periods of time. These long-term sites, both inside and outside protected areas, will continue to provide valuable information for biodiversity conservation and management.

If inventorying of the world's biota is to make the huge steps forward that are needed, then the world of systematics must change in many ways. Priorities need to be made as to which taxa are to be the focus of new and increased effort by systematists. Collection of data from museums and herbaria need to be accessed and computerised. Greater efforts need to be made towards revisionary studies rather than the description of single species.

## References

Avise, J.C. 1994 *Molecular Markers, Natural History and Evolution*. Chapman & Hall, New York.
Beddington, J. 1995 The primary requirements. *Nature* **374**: 213–14.
Chernoff, B. 1986 Systematics and long-range ecological research. In: Ke Chung Kim, K. and Knutson, L. (eds.), *Foundations for a National Biological Survey*, Association of Systematic Collections, Lawrence, Kansas, pp. 29–45.

Clark, C.W. 1981 Bioeconomics. In: May, R.M. (ed.), *Theoretical Ecology: Principles and Applications*, 2nd edn., Blackwell Scientific Publications, London, pp. 387–418.

Driml, S 1994 *Protection for Profit. Economic and financial values of the Great Barrier Reef World Heritage Area and other protected areas.* Great Barrier Reef Marine Park Authority Research Publication No 35.

Forman, R.T.T. & Gordon, M. 1986 *Landscape Ecology.* John Wiley, New York.

Fry, G.L., Norris, S., Gjelland, M. & Dahle, E. 1992 The use of geographic information systems in National Park Management: the Rondane National Park case study. In: Wilson, J.H.M., Bondrup-Nielson, S., Drysdale, C, Herman, T.B., Munro, N.W.P. and Pollock, T.L. (eds.), *Science and the Management of Protected Areas*, Elsevier, London, pp. 381–4.

Getz, W.M. & Haight, R.G. 1989 *Population harvesting: population models of fish, forest and animal resources. Monographs in Population Biology No. 27.* Princeton University Press, Princeton.

Glowka, L., Burhenne-Guilmin, F. & Synge, H. 1994 *A Guide to the Convention on Biological Diversity*, IUCN, Gland, Switzerland.

Hale, P. & Lamb D. (eds.) 1997 *Conservation outside Nature Reserves.* Centre for Conservation Biology, The University of Queensland, Brisbane.

Harris, L.R. 1984 *The Fragmented Forest.* University of Chicago Press, Chicago.

Hellawell, J.M. 1991 Development of a rationale for monitoring. In: Goldsmith, F.B. (ed.), *Monitoring for Conservation and Ecology.* Chapman & Hall, London, pp. 1–14.

Heyer, W.R., Donnelly, M.A., McDiarmid, R.W., Hayek, L.C. & Foster, M.S. (eds.) 1994 *Measuring and Monitoring Biological Diversity: Standard methods for amphibians.* Smithsonian Institution Press, Washington D.C.

Hillis, D.M. & Moritz, C. 1990 *Molecular Systematics.* Sinauer Associates, Sunderland, Massachusetts.

IUCN 1994 *IUCN Red List Categories.* IUCN, Gland, Switzerland.

Kitching, R.L., 1993 Rainforest Canopy arthropods: problems for rapid biodiversity assessment. In: *Rapid Biodiversity Assessment.* Proceedings of a Workshop, Unit for Biodiversity and Bioresources, Macquarie University, pp. 26–30.

Kitching, R.L., Orr, A.G., Thalib, L., Mitchell, H., Hopkins, M.S. & Graham, A.W. 2000 Moth assemblages as indicators of environmental quality in remnants of upland Australian rain forest. *Journal of Applied Ecology*, **37**: 284–97.

Lawton, J.D., Bignell, D.E., Bolton, B., Bloemers, G.F., Eggleton, P., Hammond,

P.H., Hodda, M.E., Holt, R.D., Larsen, T.B., Mawdsley, N.A., Stork, N. E., Srivastava, D. & Watt, A.D. 1998 Biodiversity inventories, indicator taxa and effects of habitat modification in tropical forest. *Nature* **391**: 72–6.

Lucas, P.H.C. 1992 *Protected Landscapes: A guide for policy makers and planners.* Chapman & Hall, London.

Machilis, G.E. & Tichnell, D.L. 1985 *The State of the World's Parks.* Westview Press, Boulder, Colorado.

Mound, L.A.M. & Gaston, K.J. 1993 Conservation and systematics – the agony and the ecstasy. In: Gaston, K.J., New, T.R. and Samways, M.J. (eds.), *Perspectives on Insect Conservation.* Intercept, Andover, UK, pp. 185–95.

Murray, M.G., Green, M.J.B. & Walter, K.S. 1992 *Status of Plant and Animal Inventories for Protected Areas in the Tropics.* A contribution to the ODA Strategy Programme for Research on Forestry and Agroforestry implemented by the Oxford Forestry Institute under its Forestry Research Programme. WCMC, unpublished report.

NRC 1993 *A biological survey for the Nation.* National Academy Press, Washington D.C.

Pain, S 1994 Witness for the deceased. *New Scientist* **125 (1699)**: 42–55.

Pauly, D. & Christensen, V. 1995 Primary production required to sustain global fisheries. *Nature* **374**: 255–257.

Pearson, D.L. & Cassola, F. 1992 World-wide species richness patterns of tiger beetles (Coleoptera: Cicindelidae); Indicator taxon for biodiversity and conservation studies. *Conservation Biology* 6: 376–391.

Reid, W.V., Laird, S.A., Elmez, R.G., Sittenfeld, A., Janzen, D.H., Gollin, M.A., & Juma, G. (eds.) 1993 *Biodiversity Bioprospecting.* World Resources Institute, Washington D.C.

Royama, T. 1992 *Analytical Population Dynamics.* Chapman & Hall, London.

Samways, M.J., Stork, N.E., Cracraft J., Eeley H.A.C., Foster M., Lund G., & Hilton-Taylor, C. 1995 Inventorying and Monitoring. Chapter 7.2 Scales, planning and approaches to inventorying and monitoring. In: *UNEP Global Biodiversity Assessment.* Cambridge University Press, Cambridge, pp. 475–517.

Sherman, K. 1994 Sustainability, biomass yields, and health of coastal ecosystems: An ecological perspective. *Marine Ecological Research Series* **112**: 277–301.

Solbrig, O.T. (ed.) 1991 *From Genes to Ecosystems: A research agenda for biodiversity.* International Union of Biological Sciences, Paris.

Soulé, M.E. & Kohm, K.A. (eds.) 1989 *Research Priorities for Conservation Biology.* Island Press, Washington D.C.

Spellerberg, I. 1991 *Monitoring Ecological Change.* Cambridge University Press, Cambridge.

Stohlgren, T.J., Quinn, J.F., Ruggiero, M. & Waggoner, G.S. 1995 Status of biotic inventories in US national parks. *Biological Conservation* **71**: 97–106.

Stork, N.E., & Samways M.J. (eds.) 1995a Inventorying and Monitoring. In: *UNEP. Global Biodiversity Assessment.* Cambridge University Press, Cambridge, pp. 453–543.

Stork, N.E., & Samways M.J. (eds.) 1995b Inventorying and Monitoring. Chapter 7.0 Introduction. In: *UNEP, Global Biodiversity Assessment.* Cambridge University Press, Cambridge, pp. 459–61.

Stork, N.E., & Sherman K. 1995 Inventorying and Monitoring. Chapter 7.3 Integrated approaches. In: *UNEP Global Biodiversity Assessment.* Cambridge University Press, Cambridge, pp. 517–38.

Stork, N.E 1995 Inventorying and Monitoring. Chapter 7.4 Capacity building. In: *UNEP Global Biodiversity Assessment.* Cambridge University Press, Cambridge, pp. 538–43.

UNEP 1993 *Guidelines for Country Studies on Biological Diversity.* UNEP, Nairobi.

UNEP 1995 *Global Biodiversity Assessment.* Cambridge University Press, Cambridge.

Vane-Wright, R.I. 1993 Systematics and the conservation of biodiversity: Global, national and local perspectives. In: Gaston, K.J., New, T.R. & Samways, M.J. (eds.), *Perspectives in Insect Conservation.* Intercept, Andover, pp.197–211.

Walters, C. 1986 *Adaptive Management of Renewable Resources.* Macmillan, New York.

Williams, P.H., Gaston, K.J., & Humphries, C.J. 1995 Do conservationists and molecular biologists value the differences between organisms in the same way? *Biodiversity Letters* **2**: 67–78.

# Chapter editors

Nigel Stork, James Cook University, Cairns
Tohru Nakashizuka, Research Institute for Humanity and Nature, Kyoto

# Chapter 2: Forest Ecosystems

*Chapter editors: M. J. Toda & R. L. Kitching*

## 2.1 Introduction

### 2.1.1 Forest ecosystems and IBOY

Forest ecosystems harbour the richest biodiversity on the land and, with coral reefs, are the most diverse ecosystems on earth. Both forests and reefs are characterized by the presence of 'skeleton' organisms: trees and corals respectively. These skeleton organisms play key roles in promoting and maintaining high levels of biodiversity, providing various resources such as food, nesting sites, places to hide from enemies and display sites. In addition they act as a buffer to ameliorate the physical environment for the other organisms living there.

Organisms are not distributed evenly on the surface of the earth. The most obvious global pattern in diversity is the decrease from the tropics to the polar regions. The forest ecosystems of the Western Pacific and Asia originally presented an almost continuous forest cover stretching from the taiga of eastern Siberia to the forests of eastern Australia uninterrupted by arid zones or major oceans. This presents an unparalleled opportunity to understand the dynamics of diversity along a vast latitudinal transect. Within the stated goals of the International Biodiversity Observation Year (IBOY) we will:

- establish a set of forest reference sites along this latitudinal gradient;
- establish a baseline measure of the biodiversity in each site against which to monitor the impact of global climatic changes;
- use the sites and the information collected to generate hypotheses about the relationships between biodiversity and the functioning of the forest ecosystem; and
- create a well documented set of sites for future biodiversity research and the development of its sustainable use.

Forest organisms, especially animals, are greatly diverse both taxonomically and ecologically. They occupy a wide range of microhabitats, consume many kinds of food resources and are active in different seasons and/or different times of day. Surveys carried out as part of IBOY, should focus on major compartments of the forest ecosystem in relation to important functions. A forest ecosystem can be divided in general into an above-ground space and an on- or under-ground space.

Above ground level, the leaves, flowers, fruits and seeds of trees are distributed unevenly in three-dimensional space and change temporally in both quality and amount. Microclimate changes spatio-temporally, following changes in the three-dimensional distribution of plant biomass. Within this space, many animals interact with plants through processes such as herbivory, pollination, seed dispersal, decomposition and so forth. Such animal-plant interactions will affect ecosystem functions such as primary production and reproduction of plants, as well as promoting and maintaining the diversity of both the plants and the animals themselves. In surveying forest biodiversity we need to pay attention to those organisms that spend their entire lives on or beneath plant surfaces and those which move more or less freely among individual plants. These two components of the fauna need to be sampled separately.

Organisms within soil and litter are primarily involved in the processes of decomposition leading to the cycling of nutrients and form the basis for the nutritional health of the forest ecosystem as a whole.

## 2.1.2 Secondary transects

Further survey along a series of additional gradients are also possible although the primary north-south transect should be paramount in our present endeavours. Some obvious gradients for further survey include west to east gradients from the continent out into the Pacific islands (cf. Chapter 5 of this book and Mueller-Dombois 1998), altitudinal gradients, gradients reflecting the strength of human disturbances and gradients against changes in physical environmental factors such as the amount of precipitation, soil types and so on. Most of the methods described in this manual will be equally applicable to studies on these additional gradients in the future. Limitations in resources may require a subset of survey methods be selected but as long as each method is applied in a consistent manner comparable results should emerge.

## 2.1.3 General rationale and the goals of surveys

Our basic premise in designing a sampling protocol is that a complete inventory of any particular site is, practically speaking, impossible. We focus, therefore, on producing a comparative procedure which, when applied in an identical fashion to different sites or different seasons at the same site, will allow statistical comparisons among data sets. The effort involved in application of the protocol can be increased or decreased in response to available resources or to tackle particular, more focused, ecological questions. So, for example, even though we advocate the simultaneous use of eight different devices for general arthropod sampling, any subset of these, applied in the fashion we advocate, will produce data sets comparable across sites or times. Similarly we are only too aware that additional sampling methods would target other segments of forest community. Addition of extra activities within the overall protocol, or substituting for sampling methods less relevant to the particular ecological question is of course a matter of choice by particular researchers. We do strongly advocate, however, that any overlap with the methods to be chosen be carried out in such a fashion that comparable results are obtained.

The protocol can or should be modified to produce useful results ranging from simple lists of taxa through to more encyclopedic outcomes that include detailed information on the properties of each taxon.

This is a sequential process that can be viewed as:

(A) *Pure inventory*. For the purpose of comparisons at a large geographical scale (country to country, latitude to latitude) we should choose a small number of sites where the biota are expected to be highly diversified and/or well representative of the biodiversity of the region. If we choose to expand this to enable us to detect heterogeneity at the regional scale then a range of sites should be chosen that represent different local biota within the region. The more often we repeat a survey the greater the fraction of the whole regional biota we will encounter and the more confidence we can have in extrapolating our results to estimate the total regional species richness. Results obtained through such 'pure' inventories are crucial for taxonomic and biogeographical studies and, also, provide a basis for the formulation of conservation plans.

(B) *Dynamic inventory with monitoring*. When a protocol is applied repeatedly we can monitor temporal trends in populations and communities. Such trends are vital in the formulation of conservation policy. They allow estimates of seasonality, the presence or absence of very rare species, the extinction risk of monitored taxa and the evaluation of management effectiveness.

(C) *Examining the relationships between biodiversity and environmental factors*. Biodiversity in a local area is influenced by microclimate, vegetation type, degree of disturbance, spatial structure of habitat components and so on. Within resource limitations we may establish plots that reflect these differences and can analyze the relative

*Table 2.1: Summary of required number of sites, required frequency and required period in the three types of protocol with different purposes.*

| | Purpose | Number of sites | Frequency | Period |
|---|---|---|---|---|
| Protocol A | Inventory | One or more regions (one or a few points in a region) | Rather low (or irregular) | Arbitrary or until the number of species approaches saturation |
| Protocol B | Inventory + monitoring population and community temporal change | At least one fixed point in a target area | High (in most cases, regularly recurrent) | Long-term |
| Protocol C | Inventory + analysis of microenvironmental heterogeneity affecting biodiversity | Many points (at which traps are set almost simultaneously) in a local area | A few times (sometimes once) for one local area | Short (sometimes one) |

importance of each factor. Surveys for this purpose require the sampling of many sites near simultaneously rather than repetitive surveys at particular sites. They place heavy demands on available resources of people and material for short periods of time. Results obtained from such local spatial surveys aid decision-making in diversity conservation concerning questions such as the optimum size of reserves, what vegetation types should be protected as a priority and to what degree human disturbance may be acceptable.

The adoption of particular protocols will reflect the ecological questions being posed and the resource bases of the researchers. Ideally a combination of protocols A, B and C will be most profitable providing inter- and intra-regional comparative data and information about spatial and dynamic aspects of communities and their constituent species populations within a local area. To achieve all of these goals simultaneously will require vast labour resources and may not be practically achievable. Hard decisions must therefore be made about the range of questions we will attempt to answer. Table 2.1 summarizes the protocols for particular purposes, the likely number of sites and the frequency of surveys in each case.

## 2.1.4 Core and satellite sites

For the basic IBOY-DIWPA survey all sites used should be either 'core' or 'satellite' sites depending on the completeness of the census carried out. Table 2.2 summarizes the requirements of the census for each class of site.

*Table 2.2: Components of the census for IBOY in core and satellite sites.*

| Targets | Components | Sampling methods | satellite | core |
|---|---|---|---|---|
| **Environmental variables:** | | | | |
| Precipitation and temperature | | Particular equipment | A | A |
| Light environment | | | – | A |
| **Plants:** | | | | |
| Tall-trees | with dbh > 10 cm | | A | A, B |
| Other plants | | (Regular arrangement of sub-quadrats) | – | A |
| Litter supply and decomposition rates | | (Regular arrangement of litter traps) | – | A, B |
| Phenology | | Field observation | – | (A, B) |
| **Animals:** | | | | |
| Invertebrates (general) *Beetles, spiders, mites* and etc. | Above-ground (flying) | Light trap | (A) | A, B, (C) |
| | | Malaise trap | (A) | A, B, (C) |
| | | Modified window trap | (A) | A, B, (C) |
| | Above-ground (non-flying) | Canopy knockdown | (A) | A, B, (C) |
| | | Bark spraying | (A) | A, B, (C) |
| | On- and under-ground | Pitfall trap | (A) | A, B, (C) |
| | | Litter and soil extraction | (A) | A, B, (C) |
| Invertebrates (special groups) | *Ants, termites* and etc. | Particular methods for respective group | (A) | A, B, (C) |
| Vertebrates | *Birds, bats* and etc. | Particular methods for respective group | (A) | A, B, (C) |

A: inventory, B: temporal changes (in population and community), and C: variation along microenvironment. Figures in parentheses indicate options.

In satellite sites, the main objective is to make an inventory; using methods and target taxa from among those described in this manual. The sampling methods are generally simple and thus a wide range of data may be collected. Such data can be used to detect biodiversity pattern in forests in a wide range of environments.

More comprehensive studies will be conducted in core sites. In principle, within these sites all the targets listed in Table 2.2 should be investigated. The main foci are to understand (1) temporal changes in both populations and communities, (2) variation across microenvironments and (3) interactions among the target taxa and associated environmental variables. Core sites should, where possible, have canopy access facilities (e.g. a tower system, canopy crane or perched walkway) to provide full access to biological and environmental information in three-dimensional space.

For both core and satellite sites, the total area of the study site should be, in general, at least 1 hectare or more. In a few cases the tree species diversity of a site may be so simple that smaller plots may suffice (e.g. in boreal regions). Where possible established study sites for which tree census data already exist are to be preferred so that, through repeated surveys, rapid inferences about vegetation dynamics can be made. Within core sites, a 1 ha area within which intensive census is carried out may be surrounded by less intensively studied zones.

## 2.2 Selection and establishment of observation sites

### 2.2.1 Site selection

This manual presents a detailed account of our approach, equipment and methods, covering each stage of the survey process, together with brief accounts of post-survey activities such as long-term treatment of specimens and statistical analyses of the data. It is designed to be used by those with little formal biological training. The descriptions, in general, relate to the establishment of square one-hectare plots (100 m × 100 m) but can readily be modified for application to smaller or larger sites.

As indicated, sites that represent more or less undisturbed forest are preferable for the assessment of baseline biodiversity. There are, however, some general principles involved in site selection that are important whenever the protocol is applied.

### Accessibility

The biodiversity survey of a forest plot involves a substantial amount of equipment, none of which is very large but some of which can be awkward to carry long distances. The awkwardness is exacerbated if travel involves traversing steep country, crossing major rivers or is through dense secondary vegetation. Accordingly, sites where there is access along a four-wheel drive road or by boat to within a kilometer or so of the plot should be sought. Existing tracks into the plot should also be used wherever possible or, alternatively, a single well marked track should be established, following land contours as much as is feasible. These actions reduce the magnitude of the environmental impact caused by repeated visits by a large team of researchers.

**Adjacency to laboratory facilities**
Where field teams are to be utilized to sort collected material, a simple laboratory will need to be set up within easy walking or driving distance from the plot. Depending upon the location of the plot, you may have the luxury of using existing field stations set up for research purposes. At other times, however, it may be necessary to move trestles and microscopes into the lounges of hotels, unused storage sheds or cabins on camp sites and establish a laboratory there. It is not recommended to set up such a facility under canvass, although tents for living accommodation can surround some more permanent structure such as a laboratory. Electricity is essential for such a laboratory, either through attachment to a public grid, through connection to a hotel or field station generator, or *in extremis,* by the use of smaller portable generators. This last option is to be avoided as much as possible, involving as it does the need for generator upkeep, fuelling and so forth. Whatever the arrangement, we recommend that the laboratory be no more than an hour's travel time (by whatever means) to the field site.

**Environmental uniformity**
Any hectare of forest will contain light gaps, patches of old secondary forest, minor ridge tops and drainages with or without streams in them. These variations are both inevitable and part of the forest being studied. In general, plots that contain major soil boundaries or that span waterways or forest pools should be avoided. These obstacles can usually be foreseen by prior examination of topographical and pedological maps (where these exist).

**Topography**
The ideal site is flat to enable ease of movement throughout the study area, but unfortunately, this is seldom the case. Intact forests are often reduced to small fragments of land, which are unsuitable for farming or pasture. As a result, many of these remnants are found on steep land and within gullies. This makes the survey of vegetation in particular much more difficult and increases the risk of undesirable environmental impacts upon the landscape. Nevertheless, where there is choice available, selection of flatter sites is advocated.

## 2.2.2 Plot establishment
Once sites have been selected the survey begins by marking the corners and the centre of the selected one-hectare plot, then laying out a grid of wooden pegs at

10 m intervals. The bottom left hand corner of the plot is conventionally marked as the plot's (00,00) point. This imposes on the hectare a Cartesian coordinate system that allows any point within the plot to be specified as a four digit code (e.g. 30,60 which is 30 m from the origin on the horizontal (x) axis and 60 m from the origin along the vertical (y) axis). This process involves placing 121 pegs regularly across the plot and is the first priority in beginning a survey. A time advantage can be gained if the initial pegging-out of the plot can be completed (with a small team) before the main survey begins, as it allows vegetation and arthropod surveying to begin immediately the whole survey team is in place.

In laying out a plot, the first task is to peg out the baseline – the x axis of the plot. Conventionally, this axis runs west to east from the (00,00) origin. Getting this line straight is both important and difficult. The surveyors' method of laying the line bare over the whole 100 m simply does not work in forest as gullies, patches of secondary growth and even large trees are likely to get in the way. Working with a 50 m tape, points can be aligned by eye and by compass, thus allowing pegs to be dropped into position as each of their locations is established. Each peg can then have its x and y coordinates written upon it with a waterproof marker. One surface of the top of each peg can be painted white as a base for this labelling. Orienting all the pegs so that their white faces are parallel with the x axis is a major aid should anyone become disoriented in the forest.

Once the first 50 m of baseline is established, the centre of the plot can then be marked using the 3:4:5 rule of geometry. This is easier and more accurate than using a sighting-compass or dumpy level or theodolite in thick vegetation. More sophisticated methods don't generally increase the accuracy of measurements and any benefits are out-weighed by the disadvantages of extra weight, difficulty in positioning them on wet slopes, or by the relative complexity of their use. Once the baseline and the centre point are established and marked, the remaining fixed points within the first quarter of the plot can be added by aligning them with existing known markers around the quarter hectare's perimeter. The remaining quarters of the plot can then be marked out in similar fashion by firstly identifying the corners from the established baseline, then by in-filling the points row by row.

The typical characteristics of forest topography inevitably mean that every 10 m × 10 m section of the plot is unlikely to be exact in either squareness or equal dimensions, however, the amount of error this introduces into the results is not great and is certainly tolerable. Once the marking out of the grid is completed, any point within the plot can be located to about a ±1 m error. All posts on the boundary lines (i.e. x=0, y=0, x=100, y=100) of the plot should be marked with yellow flagging tape and the 50 m lines (i.e. x=50, y=50) both horizontally and vertically, should be marked in a similar fashion. This, along with the actual coordinates written on the posts and the north/south orientation of the white faces of the pegs, aids rapid and accurate navigation within the plot.

The corners of the plot should be located using a Global Positioning System so that it can be indicated on a standard map and its location communicated successfully to others. Where the forest canopy obscures a clear GPS reading, some nearby point where readings can be obtained should be established and the plot's (00,00) point located with respect to that point by distance and bearing.

The setting up of a plot in this fashion will take a team of approximately four people up to two days. Steep topography, heavy rain and dense vegetation all increase the difficulties considerably. Once a number of 10 m × 10 m squares are clearly marked, the vegetation survey can begin as the later stages of marking out of the plot are continued. Trap-based sampling, however, cannot begin until any randomly generated (x, y) point on the plot can be located confidently and therefore must await the completion of the plot marking process.

Canopy placement is needed for several of the recommended arthropod sampling procedures – Malaise trapping, window trapping, light trapping and insecticide knockdown. We have investigated many ways of doing this but have settled on the use of a bow and arrow as the most efficient and reliable way of placing ropes into the canopy. A 40–50 lb compound bow is used with a fishing reel attachment fastened to the front just below the arrow ledge. We modify standard heavy arrows by attaching a length of fishing 'leader' along the shaft. Fishing line is then attached to this leader using a simple swivel such that when the arrow is fired the line slides down the arrow and ends up being pulled into the canopy. Attaching the fishing line to the front of the arrow, simply causes it to tumble out of control. Finally we add a rubber stopper to the end of the arrow such that if it strikes a branch or other obstruction it will bounce off rather than remain stuck in the canopy. With a little practice lines can be dropped over almost any required branch. The fishing line is used to pull heavier line into the canopy (we use 'sash cord' for this purpose) and then this thicker line is used to pull up a rope of an appropriate weight over the branch. When joining one line to another we usually smooth over the knot with plastic tape so that it is less likely to become jammed while passing through foliage. Of course there is not always an appropriate near-horizontal branch at the randomly selected (x, y) point where we wish to place a particular trap. In this case we choose the nearest appropriate branch to the selected point, noting the new (x, y) coordinates for our records afterwards. Other methods of placing lines in the canopy include the use of sling shots, cross bows, human or simian climbers or even naval style line-throwing guns.

## 2.3 Environmental variables

A very wide range of environmental variables can be measured, potentially, at each site. These range from basic meteorological observations through to geological, topographic and pedological characteristics. Many of these require access to specialized knowledge and analytical machinery. Two basic sets of measurements, however, should be taken and demand little specialized skills or equipment.

### 2.3.1 Meteorological data

Meteorological data will be needed for both the year preceding the initial survey and the most recent 10-year-averaged data (or the longest average base that is available for some sites). Basic data include the monthly total precipitation and the monthly mean air temperature: the latter is calculated by averaging the daily mean temperature (the mean of the highest and the lowest temperature in each day). Also the monthly mean maximum and minimum temperature should be given. If data are not available for the study site itself then information from the nearest meteorological station can be used. Temperature measurements from such stations may need to be modified to take into account altitudinal differences between the meteorological station and the study site. Table 2.3 is a data sheet for such meteorological data.

### 2.3.2 Light environment

In core sites, light environment should be measured as relative illumination intensity: the rate of the intensity in a forest (at least at ground level and possibly other heights) compared to the without-canopy condition. Up to twenty measuring points should be arranged in the core area. Furthermore, if a canopy access facility is available, the measurements should be conducted at various heights along the vertical profile lines. As many different heights as practical should be selected to estimate light transmission through the canopy. It is preferable that there are ten or more repetition profiles in a site. At the sites that have a seasonal leaf-fall period, twice yearly measurements are needed. Such measurements are taken, usually, during cloudy conditions. Table 2.3 is a data sheet for such measurements.

*Table 2.3: Form for site descriptions and the census of environmental variables.*

**Site descriptions:**
Name of the site:              [ _____ ]
Location:                      [ _____
                                 _____ ]
Longitude & Latitude:          [___]° N S [___]' [___]"   [___]° E W [___]' [___]"
Elevation above sea level:     [____]–[____] m
Established year:              [ _____ ]
Plot size:                     [ _____ ] m$^2$    dimension [ _____ ]
Forest types:
   *Human impact*    ❏ primary          ❏ secondary              ❏ artificial
   *Climatic*        ❏ tropical         ❏ subtropical            ❏ warm-temperate
                     ❏ cool-temperate   ❏ sub-boreal             ❏ boreal
   *Physionomy*      ❏ evergreen        ❏ semi-deciduous         ❏ deciduous
   *Elevation*       ❏ lowland          ❏ lower-montane          ❏ uppermontane
   *Composition*     ❏ broad-leaved     ❏ coniferous             ❏ mixed
Prevailing                     ❏ tree fall        ❏ large scale blow-down  ❏ fire
disturbance types:             ❏ flooding         ❏ landslide              ❏ other [ _____ ]
Soil types *:                  [ _____ ]

\* describe by reference to the FAO/ Unesco (1990) classification

**Climatic condition:**
*The most recent 10-years averaged data*
average of [ _____ ] years   ([ _____ ]–[ _____ ])

measured:   ❏ in the site    ❏ at the closest meteorological station
            ❏ at the another site [ _____ ]

Precipitation (mm):

| | JAN | FEB | MAR | APR | MAY | JUN | JUL | AUG | SEP | OCT | NOV | DEC | Total |
|---|---|---|---|---|---|---|---|---|---|---|---|---|---|
| Mean | ___ | ___ | ___ | ___ | ___ | ___ | ___ | ___ | ___ | ___ | ___ | ___ | ___ |

Air temperature (°C):

| | JAN | FEB | MAR | APR | MAY | JUN | JUL | AUG | SEP | OCT | NOV | DEC | Avg. |
|---|---|---|---|---|---|---|---|---|---|---|---|---|---|
| Mean | ___ | ___ | ___ | ___ | ___ | ___ | ___ | ___ | ___ | ___ | ___ | ___ | ___ |
| Max. | ___ | ___ | ___ | ___ | ___ | ___ | ___ | ___ | ___ | ___ | ___ | ___ | ___ |
| Min. | ___ | ___ | ___ | ___ | ___ | ___ | ___ | ___ | ___ | ___ | ___ | ___ | ___ |

*The current year data*
measured:   ❏ in the site    ❏ at the closest meteorological station
            ❏ at another site [ _____ ]

Precipitation (mm):

| | JAN | FEB | MAR | APR | MAY | JUN | JUL | AUG | SEP | OCT | NOV | DEC | Total |
|---|---|---|---|---|---|---|---|---|---|---|---|---|---|
| Mean | ___ | ___ | ___ | ___ | ___ | ___ | ___ | ___ | ___ | ___ | ___ | ___ | ___ |

Air temperature (°C):

| | JAN | FEB | MAR | APR | MAY | JUN | JUL | AUG | SEP | OCT | NOV | DEC | Avg. |
|---|---|---|---|---|---|---|---|---|---|---|---|---|---|
| Mean | ___ | ___ | ___ | ___ | ___ | ___ | ___ | ___ | ___ | ___ | ___ | ___ | ___ |
| Max. | ___ | ___ | ___ | ___ | ___ | ___ | ___ | ___ | ___ | ___ | ___ | ___ | ___ |

*Table 2.3: (continued)*

**Light environment** (for **core sites**)
*Relative illumination intensity*
measured date: [____/____/____]
measured time: [____:____ - ____:____]

| Measuring height (m) | % | s.d. | N of measure |
|---|---|---|---|
| Ground | _____ | _____ | _____ |
| _____ | _____ | _____ | _____ |
| _____ | _____ | _____ | _____ |
| _____ | _____ | _____ | _____ |
| _____ | _____ | _____ | _____ |

# 2.4 Plant diversity

## 2.4.1 Tall-trees

**Targets**

A tall-tree census should be conducted at both core and satellite participating sites. All the data should be collected for all trees (include lianas, palms and tree ferns) with a diameter at breast height ('dbh', measured at 1.2 m) of 10 cm or larger. Where previous surveys have used smaller cut-offs data can be filtered for those trees with 10 cm or greater dbh. The dbh measurements can also be used as a surrogate for tree size and the total basal area as an index of biomass of storage within censussed sites.

**Parameters of forest dynamics**

Repeated census data are needed to estimate the parameters of forest dynamics shown below. A census interval of 5 years is recommended. The basal area gain, loss, tree mortality and recruitment rate should be calculated as follows:

Basal area gain = $[\{\Sigma(B_t - B) + \Sigma(B_r)\} / (B + B_d)]^{1/t} - 1$    [1]
Basal area loss = $[\Sigma(B_d) / (B + B_d)]^{1/t} - 1$    [2]
Tree mortality = $1 - [(N - N_d) / N]^{1/t}$    [3]
Tree recruitment rate = $[(N + N_r) / N]^{1/t} - 1$    [4]

where:

B: Initial basal area of surviving trees;

$B_t$: Basal area of surviving trees after t years;

$B_r$: Basal area of recruited trees;

$B_d$: Initial basal area of dead trees;

N: Initial tree number;
$N_d$: Number of dead trees during t years;
$N_r$: Number of recruited trees during t years.
Although the definition of 'tree' in this protocol includes a stem sprouted from below breast height, tree mortality and recruitment should be calculated on an individual (not a stem) basis.

### Diversity
The total numbers of species, genera and families are necessary to estimate tree diversity at a site. It is desirable that the numbers per 500 individuals be calculated by averaging the results of ten random samples of 500 stems selected from the data. The total list of species also should be attached. Table 2.4 provides a data sheet for the tree census.

## 2.4.2 Other components
### Targets
An inventory of the other components, such as small trees, lianas, palms, shrubs, herbs, grasses and ferns should be made at core sites. Here, these components are divided into two size classes; height of 2 m or larger (and stems with dbh smaller than 10 cm) and height smaller than 2 m.

### Diversity
The occurrence of all species within each sub-quadrats should be recorded. Twenty or more sub-quadrats should be regularly set out on the core site. Each sub-quadrat should be 25 m² (5 m × 5 m) for the components with height of 2 m or larger (and stems with dbh smaller than 10 cm) and 1 m² (1 m × 1 m) for the components with height smaller than 2 m. Then we will calculate the indices representing species diversity (such as H' and J'). Table 2.5 provides a data sheet for this type of census.

## 2.4.3 Litter supply and decomposition rate
### Supply
The supply, accumulation and decomposition of litter should be investigated at core sites. At least 10 litter-traps should be set regularly in the core area. They may be made of a fine meshed cloth and placed 1 m above the ground. For the sites where litter traps will be newly set up, 1 mm mesh size and 0.5–1 m² opening area are recommend as a standard. The contents of these traps should be collected monthly and divided into flowers, seeds, leaves, branches and other materials (after species identification, the data on flowers and seeds can be used

**Biodiversity Research Methods**

*Table 2.4: Form for the census of tall-trees (parameters of forest dynamics).*

| | | | |
|---|---|---|---|
| Mean diameter: | [ _____ ] cm | s.d [ _____ ] | |
| Sum of basal area: | [ _____ ] $m^2\ ha^{-1}$ | | |
| Census interval | [ _____ ] years | ([ _____ ]–[ _____ ]) | |
| Basal area gain: | [ _____ ] $m^2\ ha^{-1}\ yr^{-1}$ | | |
| Basal area loss: | [ _____ ] $m^2\ ha^{-1}\ yr^{-1}$ | | |
| Tree mortality: | [ _____ ] $yr^{-1}$ | | |
| Tree recruitment: | [ _____ ] % $yr^{-1}$ | | |
| Number of species | [ ____ ]/[ ____ ] ha; | [ ____ ] / 500 individuals | |
| Number of genera | [ ____ ]/[ ____ ] ha; | [ ____ ] / 500 individuals | |
| Number of families | [ ____ ]/[ ____ ] ha; | [ ____ ] / 500 individuals | |

List of species in the plot (for tall-tree)

[ _____ ] in total [ _____ ] pages

(Please copy and use this form)

| Species name | Basal area $m^2\ ha^{-1}$ (%) | Individual density $ha^{-1}$ (%) |
|---|---|---|
| _____ | _____ ( ___ ) | _____ ( ___ ) |
| _____ | _____ ( ___ ) | _____ ( ___ ) |
| _____ | _____ ( ___ ) | _____ ( ___ ) |
| _____ | _____ ( ___ ) | _____ ( ___ ) |
| _____ | _____ ( ___ ) | _____ ( ___ ) |
| _____ | _____ ( ___ ) | _____ ( ___ ) |
| _____ | _____ ( ___ ) | _____ ( ___ ) |
| _____ | _____ ( ___ ) | _____ ( ___ ) |
| _____ | _____ ( ___ ) | _____ ( ___ ) |
| _____ | _____ ( ___ ) | _____ ( ___ ) |
| _____ | _____ ( ___ ) | _____ ( ___ ) |
| _____ | _____ ( ___ ) | _____ ( ___ ) |
| _____ | _____ ( ___ ) | _____ ( ___ ) |
| _____ | _____ ( ___ ) | _____ ( ___ ) |
| _____ | _____ ( ___ ) | _____ ( ___ ) |
| _____ | _____ ( ___ ) | _____ ( ___ ) |
| _____ | _____ ( ___ ) | _____ ( ___ ) |
| _____ | _____ ( ___ ) | _____ ( ___ ) |
| _____ | _____ ( ___ ) | _____ ( ___ ) |
| _____ | _____ ( ___ ) | _____ ( ___ ) |
| _____ | _____ ( ___ ) | _____ ( ___ ) |
| _____ | _____ ( ___ ) | _____ ( ___ ) |

Nomenclature: [ _____ ]

Table 2.5: *Form for the census of the other components (for core sites).*

| | | |
|---|---|---|
| Number of species | [ _____ ] / [ _____ ] m$^2$ | |
| genera | [ _____ ] / [ _____ ] m$^2$ | |
| families | [ _____ ] / [ _____ ] m$^2$ | |
| Number of trees | [ _____ ] / [ _____ ] m$^2$ | |
| palms | [ _____ ] / [ _____ ] m$^2$ | |
| lianas | [ _____ ] / [ _____ ] m$^2$ | |
| shrubs | [ _____ ] / [ _____ ] m$^2$ | |
| herbs | [ _____ ] / [ _____ ] m$^2$ | |
| grasses | [ _____ ] / [ _____ ] m$^2$ | |

List of species in the plot (for the other components)

[ _____ ] in total [ _____ ] pages
(Please copy and use this form)

| Species name | Life form* | Frequency per [_] sub-quadrants |
|---|---|---|
| _____ | _____ ( ___ ) | _____ ( ___ ) |
| _____ | _____ ( ___ ) | _____ ( ___ ) |
| _____ | _____ ( ___ ) | _____ ( ___ ) |
| _____ | _____ ( ___ ) | _____ ( ___ ) |
| _____ | _____ ( ___ ) | _____ ( ___ ) |
| _____ | _____ ( ___ ) | _____ ( ___ ) |
| _____ | _____ ( ___ ) | _____ ( ___ ) |
| _____ | _____ ( ___ ) | _____ ( ___ ) |
| _____ | _____ ( ___ ) | _____ ( ___ ) |
| _____ | _____ ( ___ ) | _____ ( ___ ) |
| _____ | _____ ( ___ ) | _____ ( ___ ) |
| _____ | _____ ( ___ ) | _____ ( ___ ) |
| _____ | _____ ( ___ ) | _____ ( ___ ) |
| _____ | _____ ( ___ ) | _____ ( ___ ) |
| _____ | _____ ( ___ ) | _____ ( ___ ) |
| _____ | _____ ( ___ ) | _____ ( ___ ) |
| _____ | _____ ( ___ ) | _____ ( ___ ) |
| _____ | _____ ( ___ ) | _____ ( ___ ) |
| _____ | _____ ( ___ ) | _____ ( ___ ) |
| _____ | _____ ( ___ ) | _____ ( ___ ) |
| _____ | _____ ( ___ ) | _____ ( ___ ) |
| _____ | _____ ( ___ ) | _____ ( ___ ) |
| _____ | _____ ( ___ ) | _____ ( ___ ) |
| _____ | _____ ( ___ ) | _____ ( ___ ) |
| _____ | _____ ( ___ ) | _____ ( ___ ) |

* Life form: T: tree P: perm, L: liana, S: shrub, H: herb, G: grass

Nomenclature: [ _____ ]

for phenology studies: see next section). The supply of litter should be expressed as dry weight of these components.

**Accumulation**
Before setting litter-traps, all the litter in the A0 layer should be sampled from a 50 cm × 50 cm area in the neighbourhood of the traps. The sample should be divided into three detailed layers (L: litter layer, F: fermentation layer, H: humus layer), and their dry weight measured.

**Rate of decomposition**
To estimate the litter decomposition rates, litter-bags may be used, made of wire netting of 1 mm and 5 mm mesh. Four sets (for each mesh size) of litter-bags should be prepared and left close to the litter traps at the major litter-fall season of the site. The leaf litter of dominant species, collected from the litter trap, may be used to fill the litter bags after drying and weighing. One of each set of the litter-bags should be collected after 1, 3, 6 and 12 months in order to calculate decomposition rates by difference. Table 2.6 is a data sheet for litter studies.

### 2.4.4 Phenology
Investigations on phenology are optional. Data on dropped flowers and seeds collected by litter-traps (see the previous section) can be incorporated.

As an optional survey, the name of species in flower and fruit should be recorded by field observation, conducted every month. The use of a facility that enables canopy access will considerably facilitate these observations. The information on flowering and fruiting should be arranged by height layer and, where possible by mode of pollination and seed dispersal. Table 2.7 is a data form for a phenological census.

## 2.5 Animal diversity: arthropods

The arthropod fauna of forests is dispersed through all habitats of the forest. It is clearly impossible to sample all possible arthropod habitats. Accordingly a selection of techniques should be used, which target a sample of the many faunal components involved. These techniques are presented in two parts. The first section describes traps and sampling methods that target a wide range of arthropod groups. Each of these methods will favour a particular subset of the arthropods within the forest, although overlapping these subsets will be unique for each method. Each method also is assumed to be equally efficient at all

*Table 2.6: Form for the census of litter supply and decomposition rates (for core sites).*

**Accumulation:** dry weight of litter (g m$^{-2}$)

| Layer | g m$^{-2}$ | s.d. | N of points |
|---|---|---|---|
| Ao (total) | | | |
| L | | | |
| F | | | |
| H | | | |

**Supply:** dry weight of litter (g m$^{-2}$)

Number of traps used: [ _____ ]

| | JAN | FEB | MAR | APR | MAY | JUN | JUL | AUG | SEP | OCT | NOV | DEC | Total |
|---|---|---|---|---|---|---|---|---|---|---|---|---|---|
| Flowers | | | | | | | | | | | | | |
| Fruits and seeds | | | | | | | | | | | | | |
| Leaves | | | | | | | | | | | | | |
| Branches | | | | | | | | | | | | | |
| Other materials | | | | | | | | | | | | | |
| Total | | | | | | | | | | | | | |

**Decomposition rates:** % of the dry weight to the initial one

| Months after the setting | 1 mm mesh | | | 5 mm mesh | | |
|---|---|---|---|---|---|---|
| | % | s.d. | N of points | % | s.d. | N of points |
| 1 | | | | | | |
| 3 | | | | | | |
| 6 | | | | | | |
| 12 | | | | | | |
| Average (month$^{-1}$) | | | | | | |

times. Secondly a set of techniques are described, which target particular groups of arthropods (drosophilid flies, Lepidoptera, spiders, ants and termites). Some of these involve the use of further more specialized traps whereas others require the development of particular skills. The selected groups are just five of the many possible. Other groups may be added to the surveys at particular sites to reflect particular expertise or interest that may be available locally.

## 2.5.1 General trapping methods
### Light traps
Light traps have been widely used for insect surveys for many years growing out of the simple observation that candles, carbide or storm lanterns used in

## Biodiversity Research Methods

*Table 2.7: Form for the census of tree phenology (optional for core sites).*

**Number of flowering species:**

|  | JAN | FEB | MAR | APR | MAY | JUN | JUL | AUG | SEP | OCT | NOV | DEC | Total |
|---|---|---|---|---|---|---|---|---|---|---|---|---|---|
| *Height layer* | | | | | | | | | | | | | |
| ___ | — | — | — | — | — | — | — | — | — | — | — | — | — |
| ___ | — | — | — | — | — | — | — | — | — | — | — | — | — |
| ___ | — | — | — | — | — | — | — | — | — | — | — | — | — |
| ___ | — | — | — | — | — | — | — | — | — | — | — | — | — |
| *Mode of pollination* Anemophily | | | | | | | | | | | | | |
| Hydrophily | — | — | — | — | — | — | — | — | — | — | — | — | — |
| Entomophily | — | — | — | — | — | — | — | — | — | — | — | — | — |
| Ornithophily | — | — | — | — | — | — | — | — | — | — | — | — | — |
| Chiropterophily | — | — | — | — | — | — | — | — | — | — | — | — | — |
| Other, unknown | — | — | — | — | — | — | — | — | — | — | — | — | — |
| Total | — | — | — | — | — | — | — | — | — | — | — | — | — |

**Number of fruiting species:**

|  | JAN | FEB | MAR | APR | MAY | JUN | JUL | AUG | SEP | OCT | NOV | DEC | Total |
|---|---|---|---|---|---|---|---|---|---|---|---|---|---|
| *Height layer* | | | | | | | | | | | | | |
| ___ | — | — | — | — | — | — | — | — | — | — | — | — | — |
| ___ | — | — | — | — | — | — | — | — | — | — | — | — | — |
| ___ | — | — | — | — | — | — | — | — | — | — | — | — | — |
| ___ | — | — | — | — | — | — | — | — | — | — | — | — | — |
| *Mode of seed dispersal* Anemochores | | | | | | | | | | | | | |
| Ornithochores | — | — | — | — | — | — | — | — | — | — | — | — | — |
| Barachores | — | — | — | — | — | — | — | — | — | — | — | — | — |
| Other, unknown | — | — | — | — | — | — | — | — | — | — | — | — | — |
| Total | — | — | — | — | — | — | — | — | — | — | — | — | — |

earlier times attracted insects freely. The modern use of light traps began with Robinson and Robinson (1950) and Frost (1952). Many designs are available for different uses and many have been developed along the way. Light traps have been widely used for insect survey both for particular target species (Bogush 1958; Geier 1960) and for whole faunas (Taylor and Taylor 1977). They target a wide range of insects but are particularly successful for Lepidoptera and Coleoptera. Light trap catches are affected by a wide variety of environmental factors and there is a lot of literature dealing with these variables (Hollingsworth *et al.* 1961). Perhaps the most important of these is the phase of the moon (Bowden 1973; Bowden and Church 1973). Light traps operated close to the period of full moon, in general, attract fewer moths than at other times, although these may be of different species.

We recommend a commercially available design based on the so-called Pennsylvania (or Texas) trap (Frost 1957). Essentially this comprises a vertically mounted 'black' light fluorescent tube with three transparent plastic vanes mounted equidistantly around it. These vanes are shaped to fit within a funnel and the funnel of 40 mm caliber sits within a replaceable bucket in which the catch accumulates. The light operates using a 12 V gel battery of the type used for powering motorcycles. To this commercially available model we add a rain protector (actually an alloy dustbin lid) and a sandwich of wooden boards beneath which the battery can be mounted. These modifications are specifically (a) so that the trap can be used in wet to very wet conditions and (b) so that the trap and its battery can be hauled into the canopy by rope. Figure 2.1 illustrates the modified design we use. We use a block of dichlorvos™ impregnated plastic as a killing agent placing it in the bucket together with torn egg-trays to provide resting places for captured insects.

Three sets of two traps, one at the ground level and the other in the canopy, may be operated simultaneously at three randomly determined points within the hectare. The light-trap points should be selected such that no set of traps is visible from any other set. Quarterly trapping intervals will be ideal at core sites, but the frequency can be reduced at satellite sites. Traps should be run for at least three nights on each sampling event, avoiding the week around the full moon. Traps run from about 5.00 p.m. until dawn. Traps should be emptied daily throughout the trapping period and the batteries recharged every day at the field laboratory.

Light traps attract so many insects that it will require a great deal of effort to sort all of the catches. We recommend sorting first Lepidoptera and Coleoptera as the primary target taxa. If sufficient human resources and time are available, you might sort Homoptera, Heteroptera, Hymenoptera (except for ants), Psocoptera, Trichoptera and others as secondary targets and then Diptera and ants as the tertiaries. Once targeted, all specimens of each order should be removed from the catches, regardless of body size. At the end of each stage of sorting, the remaining catches should be preserved in 80 % ethanol (or in a freezer) for later attention.

Lepidoptera from these light traps are the only insects that must be curated in the field. They can be sorted according to putative morpho-species at the same time. This enables many very common moths to be discarded once they have been counted. Standard setting boards can be used and the moths spread each day. These must be stored in a portable drying cabinet for the three subsequent days and are then removed from the boards, labelled and sorted. Subsequently

## Biodiversity Research Methods

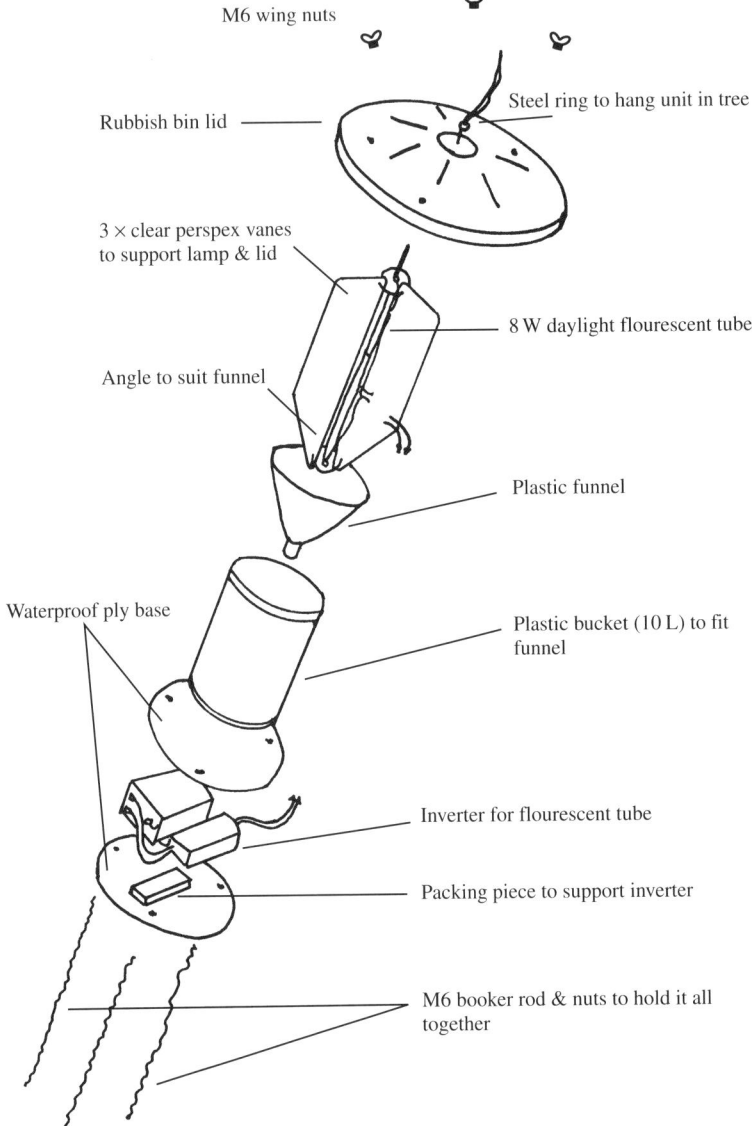

*Figure 2.1: Pennsylvania style light trap modified for rainforest use.*

they can be stored in specially constructed wooden trays within which standard museum 'unit trays' fit tightly. This preserves the material in good condition and minimizes direct handling subsequently. These processes of curation are time-consuming and require appropriate expertise. Catches could simply be layered between tissue papers in an airtight box together with an anti-fungal agent such as chlorocresol crystals. However, if there is a full team of people in the field this is not only unnecessary, but the quality of the material returned from the field is greatly enhanced. In general, the more work done in the field or field lab, the better. Coleoptera can be simply picked from the catches and preserved in 80 % ethanol for later attention. In general, light traps catch larger (as well as smaller) beetles and are undoubtedly targeting fauna which none of the other recommended methods sample efficiently.

Erecting the light traps and associated ropes is a full day's work for a team of two people. Emptying the catch each morning and opening them each evening represents about an hour's work for a team of two on each occasion. The evening work can be shortened by sending out two teams of two. The sorting, mounting and subsequent curation of the specimens will occupy three to four people full time during a two-week survey period. At least two of the helpers need to have been trained in the art of spreading moths.

Points of particular practical note when light trapping include the following.

- Returning the full buckets to the field laboratory needs to be a high priority first thing each morning so that material can be handled while fresh.
- Carefully and continuously protect mounted material from moisture, cockroaches, ants and clumsy people.
- Maintain two full sets of batteries so that one set is always fully charged. Check that batteries have 'held' their charge using a voltmeter before deploying them in the field.
- Check that each light is working and glowing steadily before leaving it in the field.
- When a light does not work, simply leave that trap for the night and service it the next day – trying to replace fluorescent tubes late in the evening, in the field and, often, in the rain, is not a good idea. Traps that have not operated for a night for whatever reason are kept open for an additional night or nights as required after the original trapping period is completed to ensure a uniform number of samples from each trap.
- Remember that the traps need to be opened late in the day every day – decide early in the day who is going to do this – it is not a popular job at the end of a hard day's work!
- Maintain effective killing bottles (four or five; these are best based on cyanide or ethyl acetate) in the laboratory for dealing with moribund material in the trap catches once they have been returned to the laboratory. Remember that these killing bottles are hazardous material.

- Take tissues to the field when emptying light traps: to close the entry at the base of the funnel before removing it and, occasionally, to dry the inside of the bucket before carrying it from the site.
- Moths and beetles are frequently perched on, but not in, the trap when they are visited each morning – sweep these into the funnel and add them to the catch.
- Number each trap clearly and check, preferably twice, that this number is retained with the catch throughout.
- In the field laboratory, traps must be tipped out into sorting trays. Sometimes catches are subdivided for ease of handling and the catches are subdivided frequently during the curation process. Ensure that at every subdivision of the catch a label is also created so that at any time in the prolonged process of curation the trap number and day of capture of every specimen can be clearly identified.

## Malaise traps

The fabric tent trap which has become known as Malaise trap was first described by Malaise (1937) and subsequently modified by a number of authors including Gressitt and Gressitt (1962) and Townes (1962). It targets free flying insects and is particularly successful in catching Diptera of which many thousands may accrue over three days of trapping. Juillet (1963) has suggested that Malaise traps are unbiased for Diptera but less so for Coleoptera and Hemiptera. Roberts (1970) disagreed with that assessment, pointing out that even catches of Diptera were influenced by both the shape and colour of the trap. To this we would add that the positioning of the trap is a crucial factor.

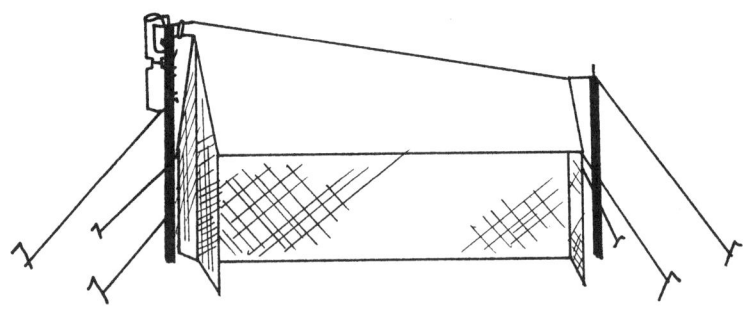

*Figure 2.2: The fully erected 'ground' Malaise trap.*

We use commercially manufactured Malaise traps that have a collector at only one end of the trap (see Figs. 2.2–2.4). We fill the collecting jar about half full with water mixed with a little bit of detergent. We have modified the basic design of the trap by adding a lightweight rectilinear frame within which the trap can be erected before it is hauled into the canopy (Fig. 2.5).

Traps deployed in the canopy catch fewer insects than those erected on the ground; the latter also contain a component of animals that simply crawl up the fabric. We deploy three sets of two traps, one at ground level and the other hauled into the canopy on rope, at three points randomly determined within the hectare. Canopy traps are manoeuvred through the lower foliage of the forest using two guide-ropes attached to opposite ends of the frame within which the trap is erected. Ideally, the interval for sampling at core sites is quarterly; the frequency can be reduced at satellite sites. Traps should be emptied daily and run for at least three days on each sampling event.

Figure 2.6 gives a sample outcome from both ground and canopy zone trapping. In Malaise-trap sampling the primary target orders are Diptera and Hymenoptera, the secondaries Coleoptera, Homoptera, Heteroptera, Psocoptera, Araneae and so on, and the tertiaries Collembola, Lepidoptera, ants and Acarina.

The erection of six traps, three of which have to be fitted within canopy frames, is a time-consuming process that requires a team of about four people one entire working day. Emptying the ground zone traps each day takes a matter of minutes but the lowering and raising of the canopy mounted traps is more time-consuming and requires a minimum of three people to 'steer' the trap into place in the tree tops.

Particular tips for operating Malaise traps include the following.
- Keep all the equipment, including tent pegs and guide ropes, necessary for a single trap in a single carry bag. Tools required – a hammer,

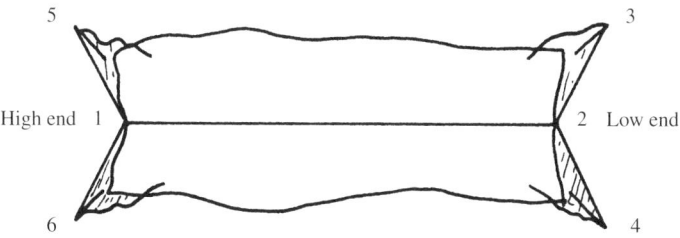

*Figure 2.3: Malaise trap 'tent' spread out to peg.*

## Biodiversity Research Methods

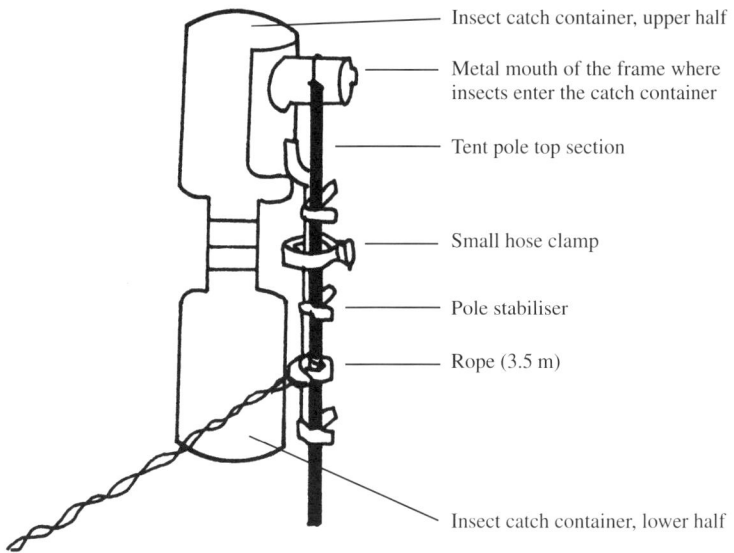

*Figure 2.4: Malaise trap catch container*

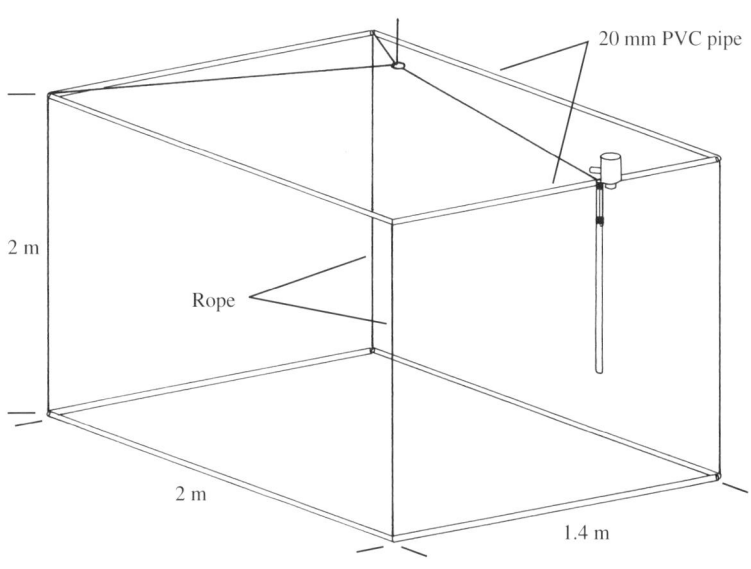

*Figure 2.5: Frame for malaise trap lifted up to forest canopy.*

screwdriver, pair of scissors and sewing kit – should be in a separate bag where they cannot tear the fabric of the traps.
- Position the trap with great care. Look for natural flyways through the ground-zone vegetation, or natural openings within the canopy.
- Look for particularly high branches from which to suspend Malaise traps – the sheer size of the traps means that the suspension points must be high to ensure proper sampling of fauna.
- Spend time and care ensuring that the traps are put up securely and well.
- Malaise traps are fragile. Repair kits, supplied with the Malaise trap, and spare parts should always be carried. Check each trap daily for damage.
- Proper labelling procedures are critical when using Malaise traps. Ensure that when the catch is transferred from the trap's catch-bottle to a sample vial, the same label or an exact duplicate is transferred with the catch.
- Number each trap in an unequivocal fashion and ensure that any numbers written on the trap bottles, for example, during previous use, do not confuse collectors.
- Wash and dry each trap immediately after use and repair any tears or other damage that may have occurred.

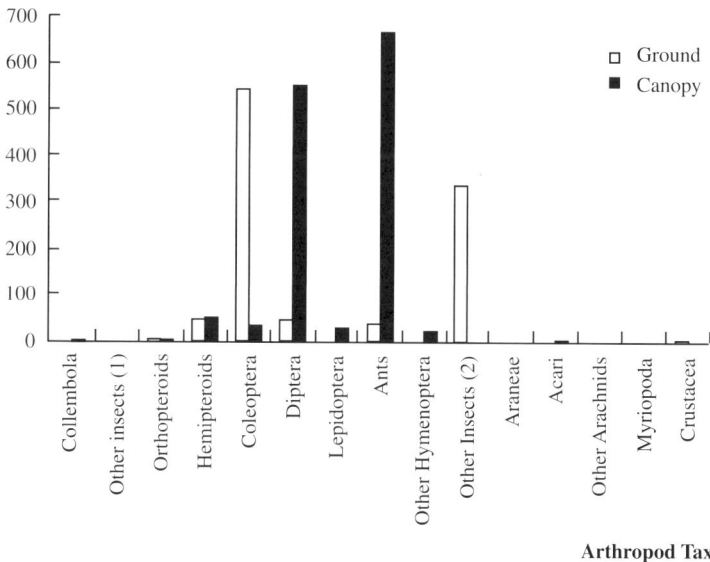

*Figure 2.6: Example of the catch from ground and canopy zones Malaise traps in tropical rainforest at Baitabag village, Madang, Papua New Guinea.*

## Window traps

Window traps work on the principle of flight interception like Malaise traps and visual attraction. They sample a wide range of flying insects, particularly stronger flying insects such as Diptera, that collide with transparent vanes (the 'windows') and drop into a collecting bucket and attract some groups of insects such as bees and Homoptera by the colour of the bucket or the reflection of the vanes. Window traps with a yellow bucket also work as yellow-pan traps and can be set at given strata of forest.

A window trap comprises a roof, two transparent vanes that intersect vertically and a yellow plastic collecting bucket (Fig. 2.7). The collecting bucket contains water to which a little detergent has been added. Small holes in the wall of the bucket allow excess water to drain following rain. Two traps, one in the canopy and the other at ground level, should be set at each of three points randomly determined within the hectare to compare insect assemblages that may occur between different forest strata. The lower traps should be suspended from low branches directly or on ropes tied horizontally between two trees above the forest floor. Higher traps are suspended by rope from a canopy pulley. Quarterly trapping intervals are ideal for core sites, but the frequency can be reduced at satellite sites. Traps should be run for at least three days on each sampling event.

When traps are set the collecting buckets are filled with two liters of water and a few drops of detergent. Traps may be emptied at daily intervals during each sampling event. On each collecting occasion the water containing the sample is poured via a funnel into a pre-labeled bottle. The trap vanes should be cleaned and the collecting buckets refilled before the traps are reset. Figure 2.8 provides a sample profile of a window trap catch from a temperate deciduous forest.

A team of two people will take a full day to suspend three traps by ropes. It will take about half an hour for two people to empty the traps and collect the samples. On return to the field laboratory samples should be filtered to concentrate them and stored in 80 % ethanol ready for ordinal level sorting.

Points to remember in operating window traps:
- Select sites with appropriate canopy gaps and horizontal branches from which to suspend the traps.
- Keep the vanes clean between collecting occasions.
- Carry the following equipment into the field: a bag to carry the bottles, a funnel, forceps, pre-labelled empty bottles with spares and bottles or carrying tanks with the replacement water supply. The spare empty bottles and the labelling tools are sometimes needed to collect samples whose volumes have been increased by rainfall.

# Forest Ecosystems

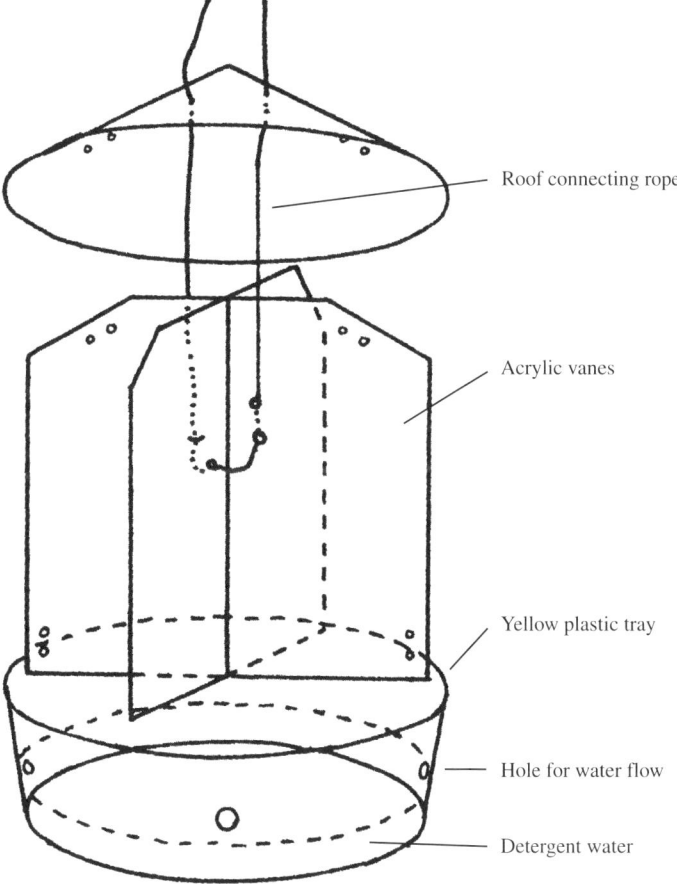

*Figure 2.7: A window trap.*

- Check the traps regularly during dry spells and top them up with water if required. If traps are emptied at close intervals this is less likely to be a problem.
- Record the numbers and details of traps that may have lost catches or had them significantly reduced by, for example, the action of storms.

## Canopy knockdown

Sampling the free-living arthropod fauna of the forest canopy using a cloud of short-lived, quick-acting pyrethrum (or pyrethroid) insecticide has become the method of choice for general canopy collecting. In a little known paper, Erwin (1990) described the history of the technique, collecting together many key references. The technique was first applied in temperate forests by Martin (1966) but opened up a new era of study of arthropods in tropical forest canopies after Roberts (1973) used a canopy fogging technique to sample canopy Orthoptera in Costa Rica (although he used the more powerful insecticide Dichlorvos™ for this purpose). Subsequently, Gagné (1979) and Erwin (1982 and subsequently) modified the technique so it could be used for the quantitative sampling of canopy faunas. Other key references include Southwood *et al.* (1982a, 1982b) and Stork (1987a, 1987b, 1988).

The technique has been used on a large scale in Australia. The basic methods in these studies are in Kitching *et al.* (1993). We recommend using a modified backpack sprayer producing an insecticidal cloud of slightly higher droplet size than the Dynafog™ machine of earlier authors. We suggest carrying out three spraying events within each one-hectare plot, each event targeting a 10 m × 10 m segment of high canopy. The centres of these sampling events must be randomly chosen and often modified to prevent overlap. A stout horizontal branch is also required in the canopy to bear the considerable weight of the spraying machine.

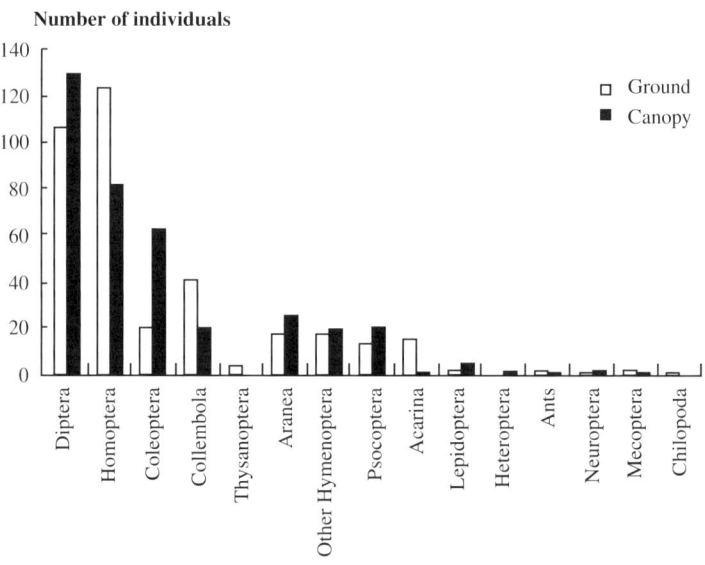

*Figure 2.8: Example of the catch from window traps in ground and canopy zones in temperate deciduous forest at Tomakomai, Hokkaido, Japan.*

At each site a rope is attached through a pulley over a high branch. A cats-cradle of lighter ropes is constructed at head height around this central suspended rope. We suggest hanging twenty 0.5 $m^2$ collecting hoops within the 10 m × 10 m plot from the cats-cradle. Each of these collecting hoops is a funnel of white plastic fabric. In the bottom of the funnel, an elasticised sleeve is sewn, into which an ethanol filled collecting vial is fitted (see Figs. 2.9 and 2.10). A pyrethroid insecticide, 'Pyrethrins 2EL™' in a Stihl™ backpack mister is used, at the concentration recommended by the manufacturer for general use. This insecticide comprises a mixture of natural pyrethrins with a piperonyl butoxide carrier. The mixture is made up with water.

Start the machine on the ground and open the spray nozzle so that a fine mist of insecticide is ejected to about 6 m distance. The sprayer is then hauled into the canopy, guided by ropes attached to its base. Deliver insecticide for five minutes at each site, during which time the machine, inevitably, spins on the rope and creates a cloud throughout the canopy above the collecting funnels. After five minutes we lower and shut off the sprayer. This procedure is conducted in the morning, as early as possible, and during a windless period. Once the spray is complete, leave the area for 30 minutes before returning to each funnel, brushing any arthropods that have fallen into them into the collecting vials. Collecting takes three to four hours after spraying. The samples

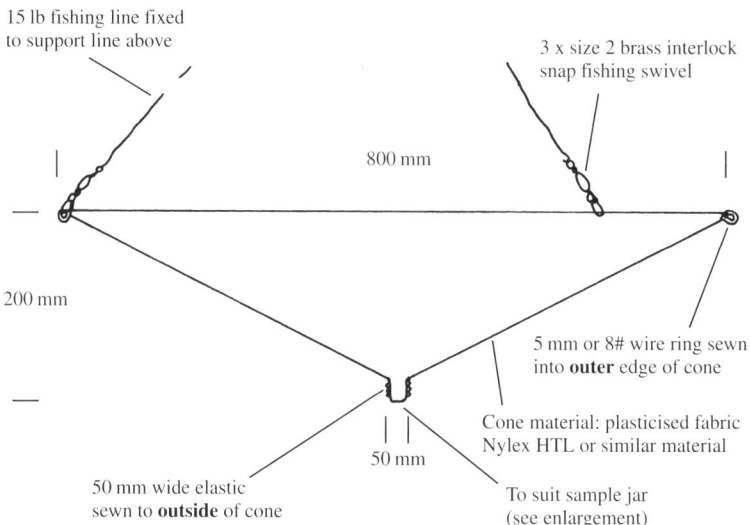

*Figure 2.9: A knockdown 'hoop' or collecting funnel.*

## Biodiversity Research Methods

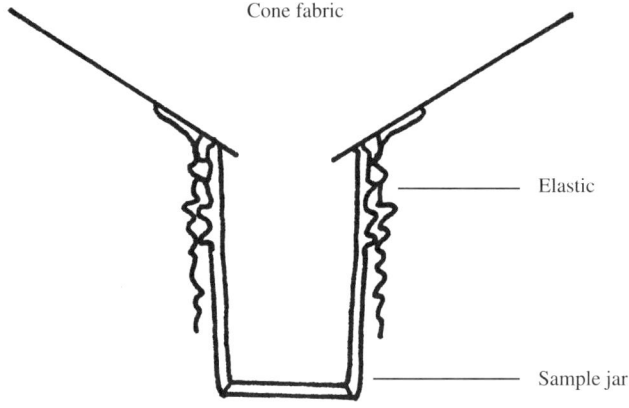

*Figure 2.10: Enlargement of sample jar holder in fogging hoop in Figure 2.9.*

are then returned to the field laboratory for sorting. Each spray event produces 20 catches for sorting but these are not themselves replicates. The catches from all twenty are combined as the sampling outcome of the spraying event. The replicates to be used for analysis are the three separate spraying events within the hectare.

Establishing a spray site once the rope is in place is an hour's work for two or three people. The actually spraying requires three people, two on the haul rope and one on the guide-rope. These three workers must wear appropriate face masks, aspirators and goggles. Tending the funnels and, finally, emptying them after three to four hours is patient and painstaking work for one person. Two sites can be sampled in one morning although we normally aim to sample one site a day in this fashion. In general, carry out spray samples late in the general arthropod survey to minimize interference between the insecticide and any other trapping method. Figure 2.11 presents a sample outcome from rainforest canopy knockdown.

Particular hints for efficient canopy sampling are as follows.
- Ensure that the branch to which the rope pulley is attached is stout and healthy – sufficient, say, to bear the weight of the heaviest member of the research team. Many kilos of operating back-pack mister filled with fuel and insecticide makes a very awkward missile should the branch give way.
- Equip each funnel with a simple clip for attaching it to the suspending ropes.
- Keep spectators, including other workers, away from the site during spraying.

# Forest Ecosystems

- Ensure that any insecticidal spillage on clothes or skin is rinsed off immediately using copious quantities of water.
- Prepare dilutions of insecticide from concentrate before taking the machine into the field.
- Carry tools and spare parts for the mister. These machines are driven by a two-stroke engine and need constant cleaning, maintenance and 'tweaking' if they are to work efficiently (be comforted, Dynafog machines are much worse!).
- Remember that as well as insecticide, fuel is required for the mister.
- Store both insecticide and fuel carefully at the field laboratory.
- Ensure that aspirators and other safety equipment are serviced regularly and that disposable filters are replaced.
- Hang collecting funnels on the cats-cradle of ropes just before the spraying event – not the night before, as these large funnels are extremely efficient collectors of rainwater and leaf litter!
- Before beginning a spray, double check that all twenty hoops have got vials fitted and that each vial is half filled with 80% ethanol.

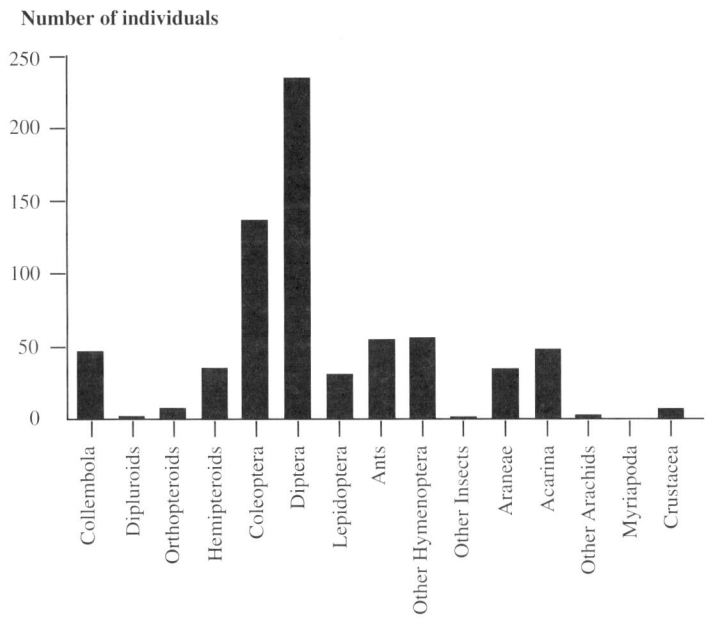

*Figure 2.11: Sample results from a canopy knockdown sampling in tropical rainforest at Robson Creek, North Queensland.*

- Ensure that all twenty vials are labelled with the number of the site and of the vial itself. Double check this!

## Bark spraying

The arthropod fauna of the bark surface of trees is rich and interesting. It is particularly rich in arachnids and beetles and its sampling adds a novel segment of the forest fauna to any survey. Bark surfaces are also easy to sample.

We recommend a simple technique that involves suspending a collecting hoop (Fig. 2.12) against a segment of tree bark at about chest height. The hoop has had its wire frame removed from about a third of its circumference (in fact we use damaged canopy spray hoops for this purpose). The hoop is pinned tightly to the bark using thumb tacks. Then mark the corners of a vertically oriented segment of bark, 1 m × 0.5 m above the edge of the collecting hoop. This segment of bark is sprayed using an aerosol can of proprietary household insecticide, again based on simple pyrethrums with piperonyl butoxide as a carrier. Spray each half square metre for about 20 seconds from a distance of about 1 metre. Holding the can any closer results in (a) the condensation of liquid insecticide on the bark and (b) the 'blowing' of insects off the bark. Over the next 30 minutes, brush down the bark gently using a camel-hair paintbrush, finally removing the catch into the plastic screw-top container placed at the base of the collecting hoop. We routinely sample thirty or forty trees per one-hectare plot in this fashion. We generally sample ten of each of the most common large trees on the plot, as identified by our vegetation survey.

Two people can comfortably carry out bark sampling procedures and should be able to sample five to six trees sequentially (depending on the availability of suitably modified collecting hoops). A team of two can sample thirty or more trees over a two-day period. Figure 2.13 is a sample result from bark spraying of one tree species.

Important tips for bark collecting include the following.
- The labels of bark collections must include the species of tree involved. Ensure this is clearly entered on each label so that those who sort the samples can read it clearly. Any abbreviations used must be self evident and standard – usually the first four letters of the genus and species (for example: ACME RESA = *Acmena resa*).
- Do not routinely sample the same aspect of each tree chosen. Take samples on different trees of the same species at different points of the compass.
- Avoid segments of bark that have foliose epiphytes on them. Encrusting epiphytes cannot be avoided but dense mats of moss or liverwort, or climbers in leaf, should be avoided.

# Forest Ecosystems

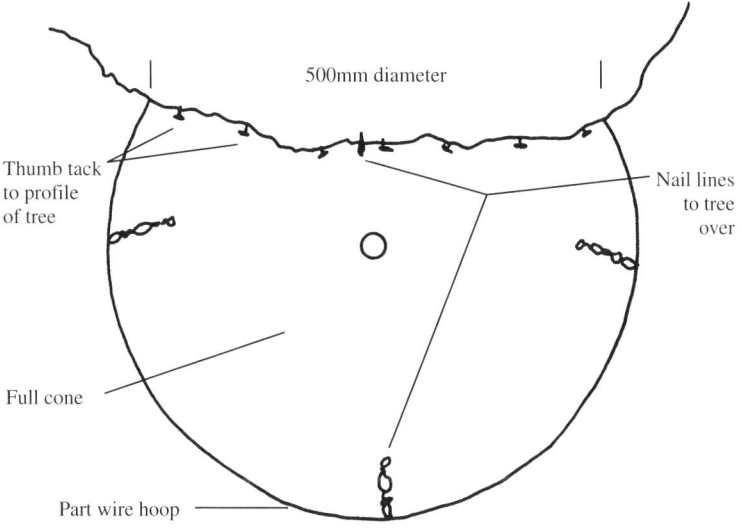

*Figure 2.12: Bark spray collecting hoop.*

- After sampling, temporarily mark each tree with surveying tape, so that it is not inadvertently re-sampled.

## Pitfall traps

Pitfall traps are one of the oldest devices known to humans, whether used to trap ants or elephants. Their use in arthropod survey is widespread, particularly when the target species are free-living ground dwelling groups. They have been used extensively for studies of spiders, Collembola, myriapods, ants and beetles. Many studies have been reported in which the capture efficiency of pitfall traps are related to factors such as weather (Mitchell 1963), available food supply (Briggs 1961), details of the placement and construction materials of the traps (Greenslade 1973) and in response to various baits (Greenslade and Greenslade 1971). Luff (1975) provides an important overview of these factors.

In designing pitfalls for general surveys, we have taken into account the material used, the ease with which they can be removed from the ground, serviced and replaced, and the need to avoid swamping either by overland water flow or direct rainfall. We use 57 mm plastic tubes as the traps. Each has perfectly smooth sides preventing escape of animals once caught.

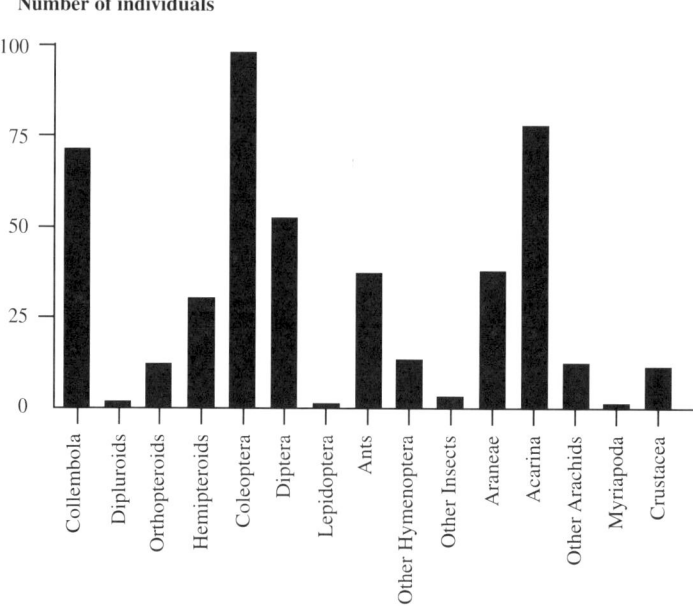

*Figure 2.13: An ordinal profile from bark spray sampling of* Knema *sp. in mixed dipterocarp forest, Kuala Belalong, Brunei.*

These fit closely inside sleeves made of tough plastic conduit (Fig. 2.14). We use five traps arranged in a $3 \times 3$ cross with 1 meter between each trap as a trapping 'unit'. The catches from each set of five are combined each day (Fig. 2.15).

Place three sets of traps centred on randomly assigned points within the hectare. Quarterly trapping intervals are ideal for core sites, but the frequency can be reduced at satellite sites. Traps should be run for three days on each sampling event and be emptied each morning during this event. A plastic roof is fitted over each to protect the catches from rain. Each tube is filled one-third full with a detergent-water mixture which is replaced each day.

Once an array of five traps is established, place a small marker of flagging tape on a wire peg next to each tube and tape off the whole area of the trap array so that others set on different tasks within the hectare will not blindly trample it.

When traps are emptied a small sieve (of the kitchen variety) is used, lined with a folded piece of very fine nylon gauze, to concentrate the catches from all five tubes in an array together. This sieving process also disposes of any rainwater that may have accumulated in traps overnight. An ethanol filled wash-

# Forest Ecosystems

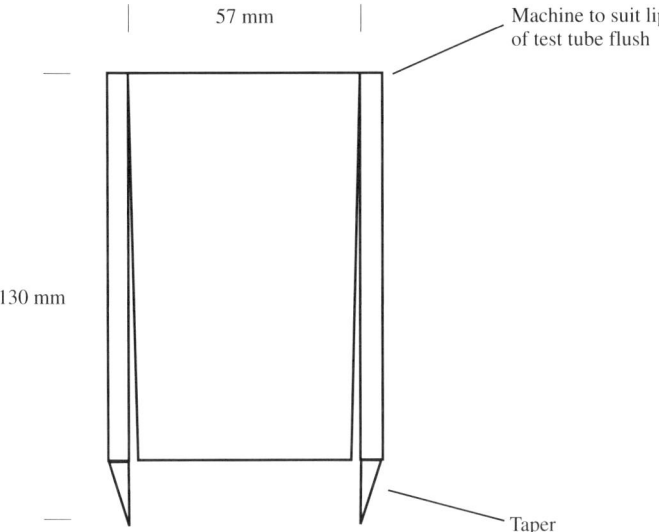

*Figure 2.14: Pitfall trap.*

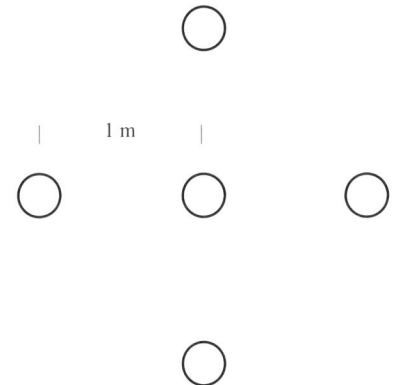

*Figure 2.15: Layout of pitfall traps; 1m between each.*

bottle is then used to wash the catch into an appropriately labelled vial. Figure 2.16 shows a characteristic ordinal profile from such pitfall surveys.

In pitfall-trap sampling the primary target orders are Coleoptera, Ants, Aranea and Isopoda, the secondaries Diptera, Orthoptera and others and the tertiaries Collembola, Acarina and Lepidoptera.

**Biodiversity Research Methods**

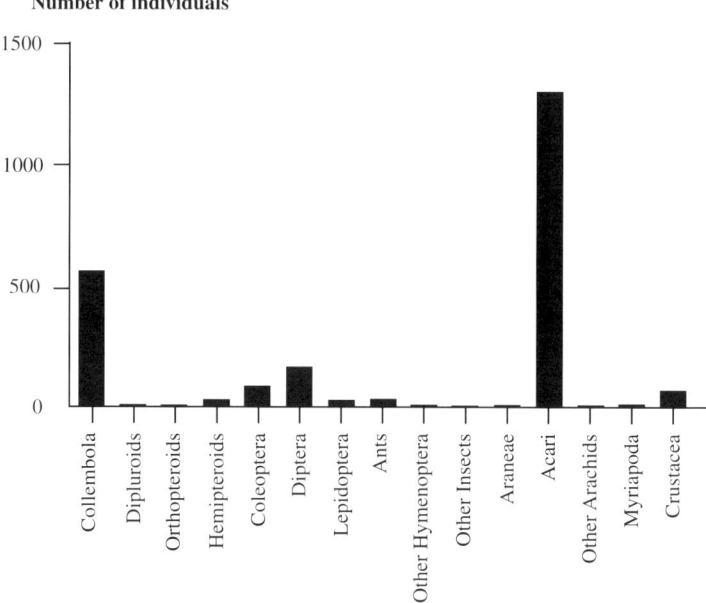

*Figure 2.16: An ordinal profile derived from pitfall trapping in subtropical rainforest in the Conondale Ranges, south-east Queensland.*

It cannot be stressed too strongly that the effectiveness of a pitfall trap depends upon how well it is positioned initially. Any discrepancy between the lip of the trap and the ground surface such that the lip protrudes will reduce significantly the number of captures. When emptying pitfalls on a daily basis it is necessary to check each day that the original perfect alignment of lip and ground is maintained when the pitfall trap is replaced in the ground. We found on one particular wet season trip in Borneo that the overland flow and consequent micro-erosion of soil was such that pitfall traps simply were not effective. No matter how well they were re-seated each time the loose soil around their rims eroded, making them ineffective. However, this was a period of exceptional wetness, even for a humid tropical rainforest.

The establishment of a set of three arrays of pitfall traps will take three people one morning. Subsequently the same number of people can service them daily in about 90 minutes. When running three sets we often set up one array per day but keep each set open for a full three days from the time of its establishment.

Below are some tips on pitfall trapping.

- Assemble all the equipment needed for a team to establish and/or service the pitfall traps together in a large plastic bag. Where a tool, such as a

hammer, is required for more than one task we strongly recommend that separate tools be purchased for each task and kept together with the equipment for that task. Scissors will still be in short supply!
- Always carry spare tubes as both glass and plastic varieties occasionally break during insertion, handling and removal.
- Attempt to keep the samples as 'clean' as possible by preventing soil particles dropping into them during removal.
- Prepare all labels before leaving the field laboratory – writing labels legibly while covered in mud during heavy rainfall is difficult.
- Clean all the pitfall equipment carefully after use. It tends to attract the most mud and dirt of any of the equipment used and is more readily cleaned at the field laboratory shortly after use, than weeks later.

## Leaf litter sampling

There are a variety of devices invented by soil zoologists (see below) by which the animals present in a volume of soil or leaf litter can be extracted, more or less efficiently, for subsequent study. Most of these are based on the observation that animals will move away from a heat source when this is applied above a mass of soil or litter. We recommend the simplest of these devices, the Tüllgren funnel. Invented in Sweden by Tüllgren in the early twentieth century, it has been much improved over the years. Modern designs are generally based upon those of Macfadyen (1955) who wrote the definitive comparative accounts of the various methods of sampling animals in soil and litter (see Macfadyen 1955, 1962). Ford (1937) was the first to use close packed arrays of Tüllgren funnels for extracting the fauna of many samples simultaneously. Paris and Pitelka (1962) discuss the many factors affecting the efficiency of use of Tüllgren funnels in describing their surveys of isopods.

Our array of funnels is contained within a custom built insect-proof box with a hinged lid and removable metal legs at each corner (Fig. 2.17). Our funnel equipment was designed by Denis Rodgers and has been constructed to be light for portability whilst being robust enough to withstand field conditions. Within this box ten 20 cm plastic funnels are inserted with their stems emerging from holes in the boxes' base. Each funnel contains a coarse mesh disk close to the base of the stem inside the funnel itself. This prevents excessive quantities of litter fragments contaminating the samples of extracted arthropods. A 20 W light bulb connected to a power source is suspended above each funnel in the box. Outside the box the funnel stems are attached to removable vials in such a way that there is no space between the stem and the vials' stoppers.

## Biodiversity Research Methods

We collect at least 10 samples, separately from those for litter accumulation, near the litter-traps (see Section 2.4.3 "Litter supply and decomposition rate" p. 40) in the one-hectare plot. Each comprises about a litre of moist leaf litter scraped up from around the selected point. We restrict our samples to the litter itself and avoid, as far as possible, including any soil. Larger branches and wood fragments are discarded. The samples are placed into sealable plastic bags in the field and then emptied into the funnels on return to the field laboratory. Extraction occurs over at least three days during which time an electric light bulb is on continually over each funnel in the array. Animals moving away from the heat source (the light bulb) pass down the stem of the funnel and are

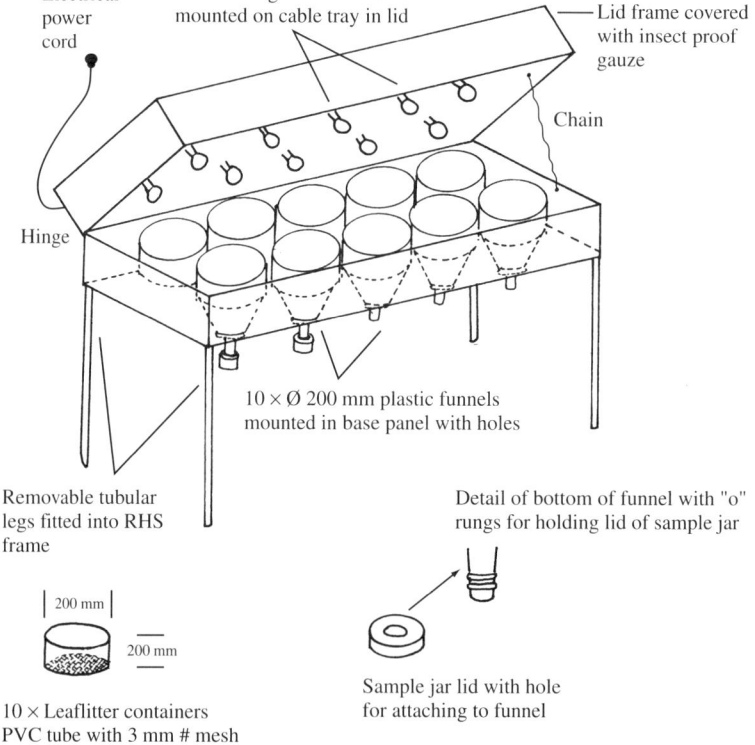

*Figure 2.17: Tullgren funnel array.*

collected in a vial of 80 % ethanol attached to its base. At the end of the extraction period the now-dry leaf litter is preserved and weighed to standardize the animal counts on a per-unit-weight basis.

Samples extracted from leaf litter in this fashion are very rich especially in Collembola and mites (see Fig. 2.18 for sample result). They take significantly longer to sort than most other samples due to the high number of arthropods present and the large amount of fine sediment in each sample.

Particular points to note in litter sampling are.
- Wear gloves when scooping litter from the forest floor.
- Ensure that a waterproof (i.e. pencil) label indicating sample number and (x, y) coordinate is placed *inside* the bag of litter.
- Ensure that an identical label is placed in the collecting vial beneath each extraction funnel as they are set up.
- Ensure that there are no gaps between the neck of the funnel and the collecting vial (this can be blocked with cotton wool if necessary) as the lights and the ethanol attract tiny insects directly, which contaminate the sample.
- Ensure that the tops of the funnels are also protected with gauze from tiny flying insects.
- When taking samples out of the funnels, ensure that an appropriate label remains with the dry leaf litter so that, once it has been weighed, the appropriate results can be reunited.
- Check the funnels regularly during the extraction process, replacing electric bulbs and topping up ethanol as required.
- Guard against fire by checking all electrical connections frequently.

## Soil sampling

Soil animals play important roles in litter decomposition and soil formation. In forest ecosystems the soil fauna ranges from nematodes to burrowing mammals although the Arthropoda are among the most apparent and diverse of the fauna. The estimation of those organisms inhabiting the adjacent leaf litter has already been discussed. To sample the soil itself we must first obtain estimates of the relative abundance of the taxa involved in a given volume of soil. Smaller soil animals inhabit the soil pores and generally cannot be directly observed. They must be extracted after soil samples have been collected in the field.

There are many extraction methods described for soil animals. The relative abundances of different taxa are best estimated following such extraction. The most popular extraction method is probably the Tüllgren funnel already described (Fig. 2.17), which extracts soil micro- and meso-arthropods efficiently.

## Biodiversity Research Methods

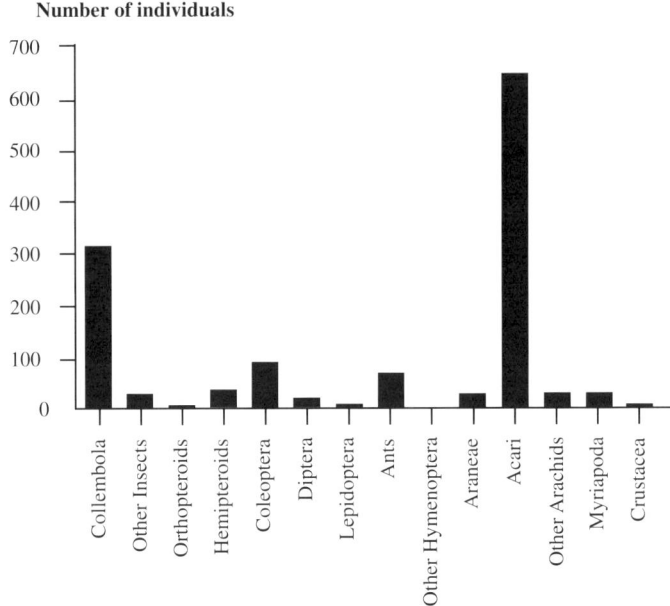

*Figure 2.18: Sample results from a set of 10 litter samples from the Kau Wildlife Area, Madang, Papua New Guinea.*

Samples for extraction are readily obtained by taking soil cores. Subsequent counts may be scaled to give numbers per square metre (or any other required surface area). We suggest that the commonly used core size of 20 cm$^2$ and 5 cm depth giving a volume of 100 ml be used for sampling the soil fauna. Soil fauna is distributed heterogeneously within any patch of forest, thus, the total number of soil cores required will vary with the degree of this heterogeneity. Only a pilot study within the target area will properly determine how many samples are required. Standard statistical formulae based on the variance of numbers within a pilot set of samples will indicate the numbers required to achieve the required levels of confidence.

Micro-arthropods are conventionally those less than 0.5 mm in length. The meso-arthropods are 0.5–5 mm in length. Both abundance and diversity of these groups may be studied following extraction from cores. The macro-soil animals (those greater than 5 mm in length) are best collected by hand.

Once collections and extractions are completed, sorting to ordinal level can be achieved using a binocular compound microscope with a low magnification.

The most abundant categories will be: Acari-Mesostigmata, Acari-Prostigmata, Acari-Astigmata, Protura and the Collembola.

The survey protocol that has been used in dry evergreen forest in Thailand provide an example that may be used as a model for other studies once appropriate sample intensities and plot size have been determined in pilot studies. Three plots of 20 m × 10 m are established randomly within the hectare. Each plot is divided into 20 subplots each 2 m × 5 m in size.

A soil pit is dug in every second subplot: a total of 10 soil pits for each 20 m × 10 m plot. Initially the top 20 cm × 20 cm × 5 cm (length × width × depth) volume of soil is removed, spread on a sheet and sorted by hand. Then the next volume down is removed and sorted and so on down to a total depth of 25 cm (5 samples deep). Arthropods are collected by aspirator or by hand and preserved in vials of 99 % ethanol.

Within each subplot a soil core sample of 20 cm$^2$ in surface area and 5 cm deep are taken for Tüllgren extraction: in total 20 cores from each plot. Extraction of each core takes at least three days although this will vary with the initial water content of the samples. Extracted organisms are stored in 99 % ethanol.

Microscopic examination follows and initial sorting to Order. Further identification requires detailed morphological examination and the use of illustrated reference works. Figure 2.19 shows a sample profile of soil animals from dry evergreen forest in Thailand.

## 2.5.2 Surveys of selected taxa
### Drosophilidae
Drosophilid flies are good candidate animals for detecting biological variability both between and within forests. Different species assemblages are likely to occur in different forest strata and in close relationship to the forest's vegetation structure. They are not efficiently sampled, however, by simple interception traps and appropriate baits must be used. We recommend a specially designed trap baited with fermented banana (Toda 1977).

The trap design is shown in Figure 2.20. It is constructed in two parts: a darkened bait chamber attached to a transparent collecting chamber. The two parts are joined simply by taping them together. Drosophilid flies enter the bait chamber through a mesh screen designed to exclude larger insects that might damage the catch. Once within the bait chamber they are attracted to the lighter, transparent end of the trap and fall through a funnel into the collecting bottle.

The bait is made by mixing soft bananas with a teaspoon of yeast. About 200 g of the bait mixture is required for each trap. Each baited trap is exposed

## Biodiversity Research Methods

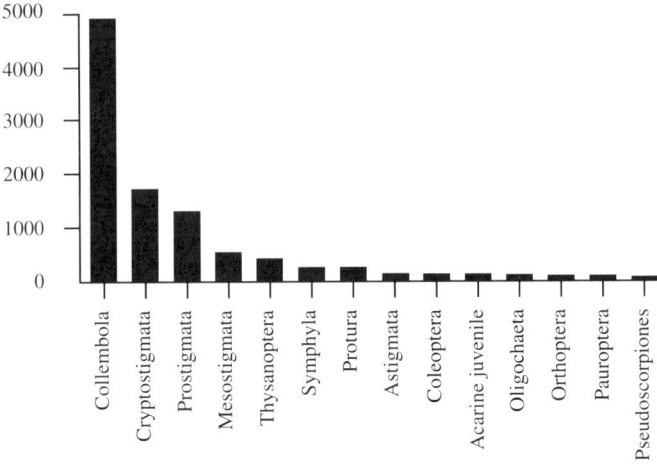

*Figure 2.19: Abundances of soil mesofauna in a dry evergreen forest in Thailand.*

in the forest for a week. At the end of this time any living or dead flies that have not entered the sample bottle may be collected using an aspirator or forceps. The trap may be reset after cleaning and re-baiting.

We suggest that traps be suspended along a rope at intervals into the forest canopy. Traps should be set at 0.5 m above ground, then at 1.5 m and then at 5 m intervals up to the canopy. Using this method, the herbaceous, shrub, sub-canopy and canopy layers of the forest can be sampled. The higher traps can all, with care, be suspended from the same rope (Fig. 2.21). The set of traps should be duplicated three times within a one-hectare plot.

It will take two people a day to set up the ropes and traps initially. Subsequently it takes about six minutes per trap to remove flies from the trap itself, change the sample bottle, clean the trap parts, replace the bait and re-assemble the trap.

Particular points to remember in using these bait traps are:
- The equipment required to operate these traps comprises bait, bait container, sampling bottles with fixative (see below), spare bottles, aspirators, forceps, labelling tools (paper, pencil and scissors), adhesive tape and tissues for trap cleaning.
- Traps should be set tilted so as to prevent the accumulation of rainfall and to increase the speed of the positive phototaxis by which flies are attracted into the collection chamber.

# Forest Ecosystems

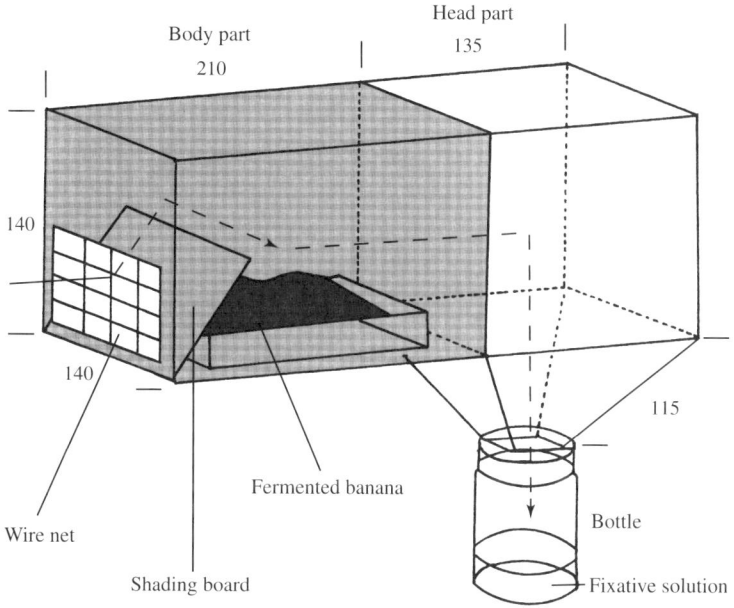

*Figure 2.20: Trap design (length unit = mm).*

- Although one week's collecting is recommended as a standard, this can be varied. After about 10 days though, specimens not in fixative begin to rot.
- In tropical and subtropical situations the rope from which the traps are suspended should be protected above and below the traps with insect repellent such as 'Stickem'™ to prevent access to the traps by ants and other marauders.
- Appropriate fixatives are 70 % ethanol, Kahleí's fluid (distilled water, 95 % ethanol, formaldehyde and glacial acetic acid in a 28:17:6:2 ratio). Where examination of internal organs is required (for population dynamics or taxonomic study) Kahleí's solution should be used. We recommend 70 % ethanol for subsequent storage of specimens regardless of the fixative in which they were collected.

## Lepidoptera

Lepidoptera, butterflies and moths, are taxonomically speaking the best known Order of insects and hence are ideal candidates for in-depth, species-level biodiversity studies at any site. That having been said it must be noted that large

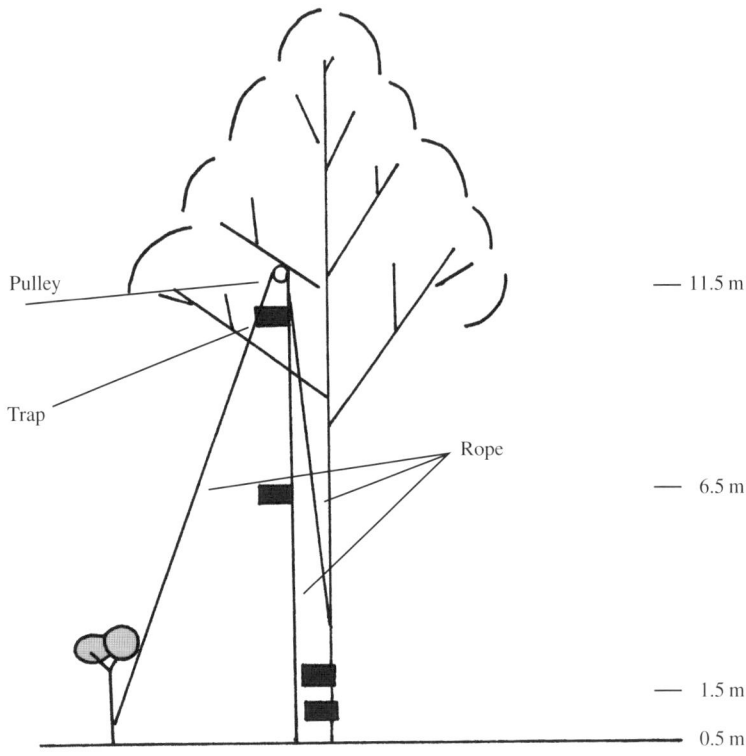

*Figure 2.21: Vertical trap setting using a rope system.*

segments of the Order Lepidoptera are not well-known – particularly the smaller moths. Following recent classifications (Common 1990, Minet 1991, Neilsen *et al.* 1996) the Order Lepidoptera is divided into six sub-orders. The larger better known species all fall within the clade Myoglossata. Within this huge taxon only the 'higher' Lepidoptera, the Ditrysia, really lend themselves to species-level biodiversity surveys. Conventionally, then the Ditrysia are subdivided into 'micro-' and 'macro-' Lepidoptera. Originally these were collectors' designations for the species that were readily treated by amateurs (the 'macros') as opposed to those that were not (the 'micros'). Brock's (1971) and Minet's (1991) cladistic analyses now identify the Macrolepidoptera as a genuine clade within the Order and, strictly speaking, it is this clade that lends itself to species-level survey.

For most of the Asian/Australasian 'green belt' region there are well accepted, illustrated and available handbooks for identification to species of the butterflies. For moths, although many major identified reference collections do exist, the situation is far less satisfactory. Good introductory handbooks are beginning to appear for moths but the identification of macro-moths still presents a much larger problem than for butterflies. We recommend that a carefully curated reference collection be established for each survey site and that, as opportunity presents itself, this collection or parts of it be matched up with national or international collections within the region concerned.

The night-flying moths are well surveyed using light traps (see above). The use of these traps can be intensified with monthly, bimonthly or quarterly samples being taken if a special focus on the Lepidoptera is to be made. Day flying Lepidoptera (usually just the butterflies) require different survey techniques. The remainder of this section will be devoted to such procedures.

Three different techniques are suggested for butterfly (and day-flying moth) surveys depending on how familiar the surveyors are with the fauna itself. In practice a combination of all three methods may be preferred.

*Standard walk*

When a butterfly fauna is very familiar to the observer then the 'standard walk' method can be used. First formalized by Pollard (Pollard *et al.* 1975; Moore 1975) this method involves walking a set route and recording all the butterflies seen within a standard distance of that route – that is: for the few metres visible on each side of a footpath, for example. The regular walk can then be repeated on a daily, weekly or monthly basis allowing the gradual build-up of a picture of the entire butterfly fauna of an area.

This method has been widely used in assessing the conservation value of particular locations in temperate and subtropical habitats (e.g. Hill 1999). It does require that the majority of butterflies be recognizable without capture (either with or without a preliminary collecting period during which a reference collection may be established).

In tropical and tall forested regions it is of limited use for several reasons:
- there may be many small and hard-to-identify species such as lycaenids and hesperiids;
- whole segments of the fauna may simply be invisible – in the canopy for example;
- some groups even of larger butterflies contain many cryptic species which are hard to identify on the wing (e.g. species of *Euploea* or *Graphium*);

- the line of walk may not include all key micro-environments within a region so that, e.g. butterflies of swampy or very steep ground may be overlooked;
- the survey walk should be carried out at the same time of day each time – in tropical situations this should be in the morning.

*Catch per unit effort*
An alternative to the 'standard walk' method for assessing butterfly diversity is to use a 'catch per unit effort' approach. Inasmuch as this involves actually collecting specimens it will have a greater impact on the fauna of an area and hence should be used with greater care, possibly more infrequently, than a simple non-collecting walk. If quantitative results are to be inferred from such a procedure then care should be taken that the actual collecting effort in each location or on each occasion should be near identical. The collecting route should be the same on each occasion – but this should involve a number of separate locations with a set collecting time at each within a particular forest. Particular attention should be paid to sites at which butterflies accumulate: mud puddles, rotting fruits and ridge or hilltops are particularly useful in this regard.

Again it will take repeated field visits to build up a picture of the fauna of a site. This method has been used effectively in tropical sites including the densest tropical locations (e.g. Tan *et al.* 1992, Orr and Haeuser 1996).

The goal of a standardised survey has several implications:
- the same number of collectors (people) should be used on each occasion;
- equipment should be standardised (size of nets, etc.);
- the skill levels of each worker should be more or less the same;
- handling time for each specimen should be standard.

*Butterfly bait trapping*
A variety of baits attract butterflies of certain groups. Most commonly used are fruit or dung baits. Rotting fruit is most easily obtained and attracts a range of larger more spectacular butterflies, many of which are seldom seen in the forest without the use of baits.

A simple butterfly trap comprises a box shaped cage of netting about 40 to 50 cm in dimension with a solid base of plastic or plywood. This can be readily constructed around a wire frame. An open slot 3 to 4 cm high along the base on two opposite sides allows access by butterflies. A bait of overripe bananas or other soft fruit is placed each day within the trap which is then suspended at about head height from a branch. Butterflies entering the trap will feed then rise within the trap and be unable to escape. Generally they will sit motionless on the cage sides until the trap is visited. They should be removed from the cage by hand each day of a survey.

*Combination survey*

The recommended combination of techniques to be used to assess butterfly diversity within a one-hectare plot are:
- a preliminary survey and collecting period of two to three days in which the larger butterflies are netted, identified and, if necessary, preserved for future reference;
- a one hour walk each day for four days carried out at about 10.00 a.m. with identification and counts made of all larger butterflies (Pieridae, Papilionidae, Nymphalidae);
- one hour's intensive collecting for Hesperiidae and Lycaenidae at each of four locations within the hectare (chosen on the basis of the preliminary survey) carried out on each of four days;
- deployment of four fruit-bait traps for four days at four random locations within the hectare.

Properly designed and weather-permitting this entire survey may be carried out by one person in four days. It may be repeated at quarterly or even monthly intervals if a more complete picture of the species diversity is required.

## Spiders

Spiders are a highly diverse and ubiquitous group of arthropods. They are universally predatory in feeding behaviour. They are not well known taxonomically but are a group in which interest is developing both for intrinsic and applied reasons. Spiders may be useful sources of pharmacological information and are potential agents for biocontrol. They clearly play an important ecological role in forests, particularly as one approaches the tropics. They are very good candidates for special study during a biodiversity survey.

Several of the general sampling techniques will yield good numbers of specimens of spiders. Canopy knockdown, bark spraying, pitfall traps and even Malaise traps will sample the more mobile fraction of the spider fauna. Many spiders, however, will not be sampled. Many groups of spiders are sit-and-wait-predators using webs or other devices to entrap their prey. Such species will not generally enter traps. Jonathon Coddington and his colleagues based at the Smithsonian Institution have developed a protocol for the estimation of this less trappable segment of the spider fauna. The following account is based almost entirely on their approach. It employs trained collectors and is essentially a catch per unit effort method such as the one described for butterflies above.

The 'Coddington Protocol' was eventually refined to four basic collecting methods following an evaluation of many others.

*Looking up*
This searching technique accesses the herb layer, shrub layer and the tree surfaces of a forest. Spiders are handpicked by the collector searching any structures above knee-height. This procedure should be carried out in one-hour units during the morning (7.00–12.00) and evening periods (19.00–24.00). The more time or collectors involved the more spiders will be collected but the catches can be reduced usefully to numbers per collecting hour. Collecting success will also increase with experience, so a preliminary period of familiarization and training by an experienced arachnologist is very useful.

*Looking down*
This collecting technique targets the leaf litter, holes in the ground, logs and rocks. It involves handpicking spiders while crawling about the forest floor on hands and knees. Leaf litter, forest floor debris and the shortest of vegetation can be searched intensively. Again this should be repeated for some hours and catches per collecting hour computed. Morning and evening collections should also be taken. Earlier comments about the role of experience clearly also apply here.

*Beating*
This technique targets spiders on the herb and shrub layer. It employs half square metre beating trays that are held under vegetation, which is tapped until no more spiders fall upon the tray. A range of vegetation species and types can be sampled in this fashion. Coddington and his colleagues recommend that 25 such 'beat collections' should be the sampling unit in any area being surveyed. Such a unit will take about one hour to complete.

*Litter sifting*
This collecting technique focuses on the leaf litter exclusively. The litter spread over two square metres of forest floor is gathered up, sifted and sorted on a white sheet from which the spiders are collected. Again a single worker can process such a sample in about an hour.

These methods can be applied within a survey plot or as plotless techniques within the forest as a whole. We recommend that each be applied within the one-hectare survey plots and that the set of techniques be repeated twice a day (as indicated) on perhaps three occasions during the course of a survey.

Spiders collected in this fashion should be preserved in 80 % ethanol. Sorting spiders to family and below is difficult and in general needs the collaboration of an experienced arachnologist. Such collaborations are best established before surveys are begun, as the prior commitment of the specialist can prevent unnecessary collecting of material that cannot or will not be sorted to produce analysable data.

## Ants

Ants are a group of great biomass, diversity and ecological importance in forests and are therefore an obvious target group for single taxon surveys. They are particularly important in tropical lowland forests where exceptional richness and abundance may be encountered. They are most commonly involved in predatory interaction with other arthropods but many mutualistic interactions involving ants also occur (Hölldobler and Wilson 1990). They have been used with success also as indicators of forest quality (e.g. Maryati 1994).

Although a complete inventory of ant species in any place is a desirable end it is generally too time-consuming a goal. We turn, instead, to comparative methods based on rapid assessment techniques. We advocate here a standardized method for use in forests, which will complement the ant collections emerging from the standard general sampling methods used in intensive surveys. Ants are abundant in the collections from all of the general collecting methods advocated above except Malaise traps. Canopy knockdown, bark sampling, litter extraction and pitfall trapping are particularly rewarding for students of the Formicidae.

*Where are the ants?*

The life styles of ants are extremely diverse. They can be found in a very large range of forest habitat components. Foraging workers may be found on the ground and vegetation wherever energy sources may be found. Key sugar (and amino-acid) resources for foraging ants are represented by nectar producing flowers, extra-floral nectaries, myrmecophilous lycaenid butterfly larvae and honeydew-producing homopteran colonies. Foraging workers also seek out proteinaceous food for the ant-brood such as insects and other arthropods. These they subdue and carry back to the nest. Some ant species are totally subterranean whereas others are cryptic, living beneath the bark of trees. Yet others live high in the forest canopy. Some species are completely nocturnal and are thus very cryptic in daylight hours. In survey work collections of several castes or forms including major (soldier) and minor workers, queens and males are particularly valuable taxonomically. Accordingly samples from nests are highly desirable.

When devising sampling techniques for ant survey knowledge of all these factors will enable a majority of ant species within any patch of forest to be surveyed. With experience additional microhabitats can be located and a more complete survey carried out.

*Handling ants*
Ants are usually hand-collected either using forceps or an aspirator. With forceps ants can be placed directly into vials of 80 % ethanol whereas aspirated specimens must be transferred after capture, usually after use of some intermediate killing agent such as ethyl acetate. Both handling methods have advantages and disadvantages. Using forceps leads to failure to catch the ant on some occasions but, on the other hand, avoids the debris and other extraneous material so frequently picked up by aspirators. When handling ants the key goal is to maintain the specimen in a taxonomically useful (that is: undamaged) condition. Ants may be picked up with forceps by the legs or (gently) by the thorax. Avoid picking them up by the fragile antennae. Very small and lightweight ants may be picked up by adhesion using ethanol-wetted forceps. Minute ants may be immobilized by sprinkling them with drops of ethanol.

*Standardized sampling methods: the quadra protocol*
A standardized sampling strategy is required if we are to compare the ant faunas across several sites. Such a standard strategy should require minimum equipment, manpower, time or expertise. Several such protocols have been advocated by myrmecologists in recent times and there are advantages and disadvantages attached to each. The so-called 'quadra-method' is designed to obtain a species-list of ants from an area quickly and efficiently. It uses four complementary methods and has been tested in both tropical and temperate forest situations. It combines the use of honey baits, soil sampling, the sifting of leaf litter and hand collection (Fig. 2.22). It needs no expensive or heavy tools and can be applied in virtually any habitat. We reiterate that this protocol will complement the collections made by a variety of traps at a survey site. The execution of the whole set of sampling methods in the quadra-protocol will take three collectors, with some experience, approximately two days.

*Honey baits* should be set out on the ground along a transect at an interval of about 3 to 4 m. Forty-five baits should be set in this fashion each comprising drops of honey placed on a small square of cotton. Another 45 baits should be placed on tree bark adjacent to the ground baits. If time permits the full set of 90 baiting samples may be repeated at night as well as during the day. The ants associated with each bait may be collected after an hour, either by simply picking up and preserving all of the ants using the cotton square or, with experience, by picking off a few individuals of different species by hand. This method will quickly allow the identification of the dominant ground and bark foraging species and allow major and minor workers of the same species to be collected simultaneously (Yamane *et al.* 1996).

*Soil samples.* Fifteen $20 \times 20 \times 10$ cm soil samples should be taken along the honey bait transect, five for every 15 baits or so. These should be sifted using a hand sieve and a flat white pan. All the individuals found on the pan should be collected. They can be handpicked and placed in ethanol immediately. The ants from each sample should be retained in separate vials, appropriately labelled. Soil sifting is the most time-consuming of the methods included in the quadra-protocol but is to be preferred over some 'automatic' extraction methods as minute moving ants are readily seen in a white pan but may be otherwise overlooked. It has proved particularly useful for detecting rare subterranean species. The method, suitably repeated, can be used for biomass and abundance estimations.

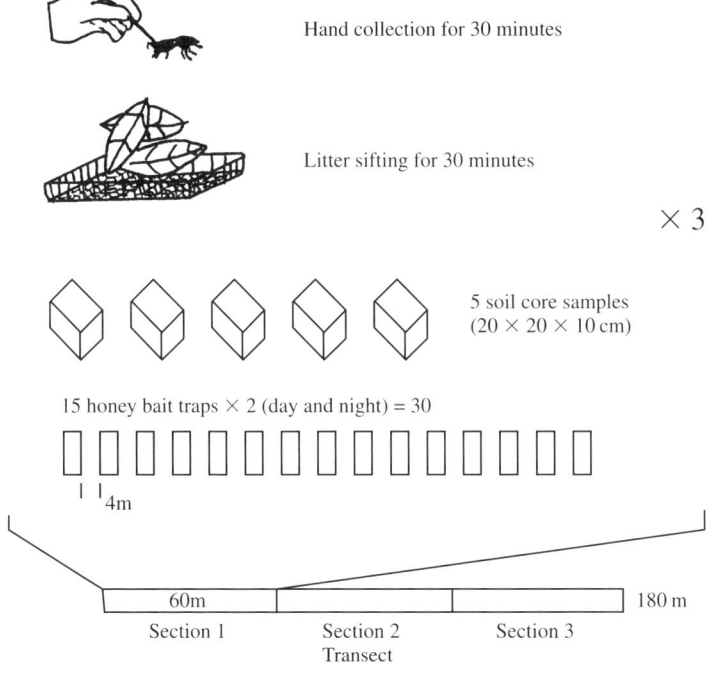

Figure 2.22: *Four collecting methods employed in 'Quadra-protocol'.*

*Leaf litter sampling.* In the three sections of the honey bait transect identified for the soil sampling exercise, leaf litter may be sifted for a standard 30 minutes by a single person with a hand sieve and white pan. Again this hand sorting method has advantages over extraction techniques and is efficient in detecting very small species. Although species found will often overlap with those from the soil each forest component does possess unique species that can be detected using these techniques.

*Hand collecting.* Lastly we advocate that in each of the three sections of the bait transect one person searches as many microhabitats as possible focussing particularly on those forest elements not otherwise sampled directly. Lower levels of vegetation such as small trees, the bases of tree trunks and decaying wood will be main targets. Tree ants of the genus *Polyrachis* will only be detected in this manner. Some species of *Crematogaster* or *Camponotus* which associate with particular plant species will also be discovered only by this general hand collecting.

*Data analyses*

Initial data analyses should be simple. An example of simple calculations following an application of the quadra-protocol in Tawau and Sayap is presented in Table 2.8. Collection efficacy can be compared across the four methods to test which may be overlooked if a more rapid assessment is required. The soil sifting samples must be retained in any combination of methods used. At least two such methods must be used (Fig. 2.23). Honey baits are less

*Table 2.8: Species number collected by 4 methods.*

|  | HC | LS | SO | HB | Total |
|---|---|---|---|---|---|
| **Tawauuuu** |  |  |  |  |  |
| S1 | 10 | 17 | 26 | 18 | 58 |
| S2 | 12 | 28 | 33 | 22 | 78 |
| S3 | 16 | 23 | 28 | 19 | 67 |
| **Total** | **26** | **47** | **63** | **35** | **123** |
| **Sayap** |  |  |  |  |  |
| S1 | 5 | 15 | 27 | 16 | 48 |
| S2 | 8 | 17 | 9 | 12 | 40 |
| S3 | 12 | 16 | 5 | 9 | 35 |
| **Total** | **18** | **33** | **36** | **20** | **77** |

HC: hand collecting, LS: litter sifting, SO: soil core sampling, HB: honey bait.

# Forest Ecosystems

*Figure 2.23: Collection of ants in the field. A: collecting litter ants with a handy sifter and white pan. B: taking soil sample. C: searching ant nests in decaying fallen tree twigs on the forest floor. D: collecting ants with a handy vacuum cleaner (this is not mentioned in the text, but often very useful for collecting quick-moving ants).*

productive than either soil or litter sifting in both forest types. Within any region, faunal comparisons across ecosystem types, altitudes and so forth are valuable, but require extensive surveying.

*Specimen maintenance and identification*
All ants collected should be kept for later identification and reference by other workers. Ant specimens can be preserved indefinitely in small vials of 80% ethanol but may be dry-mounted by pointing (Fig. 2.24). Specimens in ethanol are more resistant to the depredations of fungi, museum beetles and psocids than are dry specimens. They are however susceptible to drying up and preservative levels need to be checked frequently. Dry mounted specimens, however, are easier to identify and manipulate. The choice between the two methods depends on circumstances. Dry mounted specimens are preferred if collections can be kept under climate control.

Bolton's (1994) keys can be used to identify any ant genus although they require some specialist knowledge and experience. Identification beyond the generic level often poses problems because of the dearth of specialist monographs for many regions of the world. Further sorting must be done in collaboration with appropriate ant specialists.

**Termites**
Termites are widely distributed in tropical terrestrial ecosystems, playing an important role in the processes of dead plant decomposition and soil formation (Lee and Wood 1971; Wood and Sands 1978). They are a good indicator group for both biogeographical and ecological analyses. They are also useful in ecosystem monitoring for several reasons:
- their taxonomic and ecological diversification;
- their relatively sedentary habits;
- their taxonomic tractability;
- the fact that individuals are present year round;
- their functional importance in ecosystems;
- their predictable responses to disturbance (Brown 1995).

As many as 50 species and up to 10 000 individual termites per square metre have been reported in tropical forests (Abe and Matsumoto 1979; Eggleton *et al.* 1995). Information on termites in the western Pacific and Asian region is available. Roonwal (1970) has discussed their general biology. The composition of local faunas have been described by Ahmad (1965), Krishna (1965), Roonwal and Maiti (1966), Ikehara (1966), Chhotani (1970), Morimoto (1973), Abe (1984), Thapa (1981) and Tho (1992). Their responses to latitudinal (Abe 1987) and altitudinal (Collins 1984) changes have been discussed. Various other aspects of their ecology are dealt with by Abe (1979), Abe and Matsumoto (1979), Matsumoto and Abe (1979), Watanabe *et al.* (1984), Jones (1996, 2000) and Eggleton *et al.* (1999).

# Forest Ecosystems

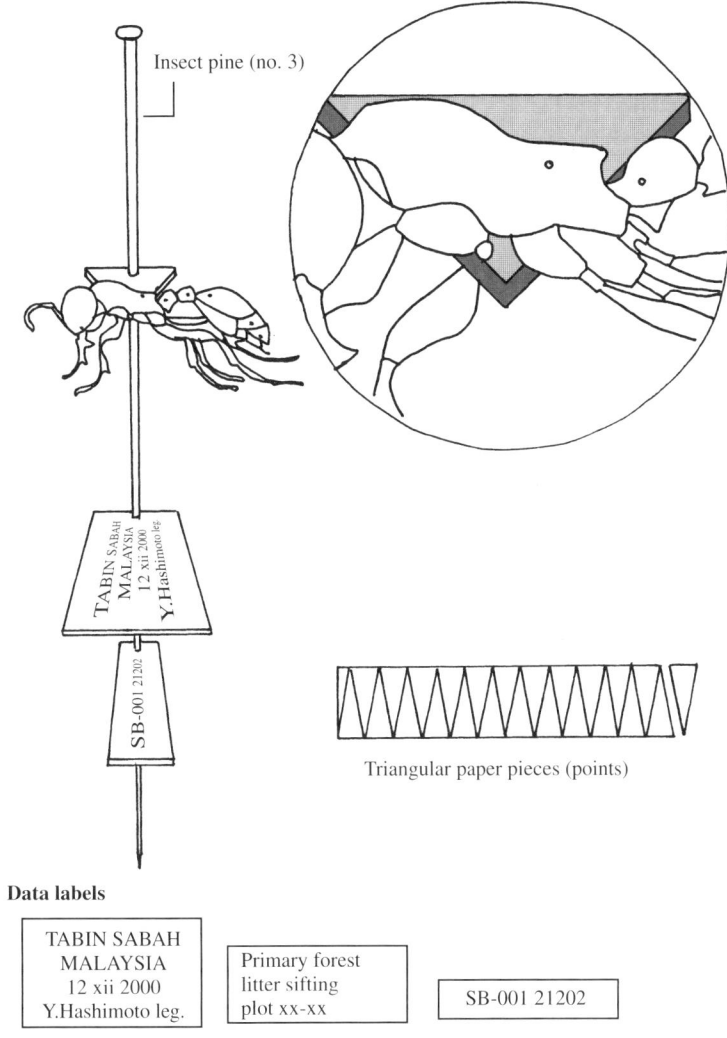

*Figure 2.24: Mounted ant specimen on a pin. Preparation of specimens of fine quality greatly helps taxonomists who cooperate with ecologists in identifying species.*

Within the framework of IBOY-DIWPA, we intend to clarify the distribution of termite species diversity in the western Pacific and Asia in relation to their ecosystem function and the impact of human modifications of ecosystems. It is essential to standardize survey methods for these purposes.

# Biodiversity Research Methods

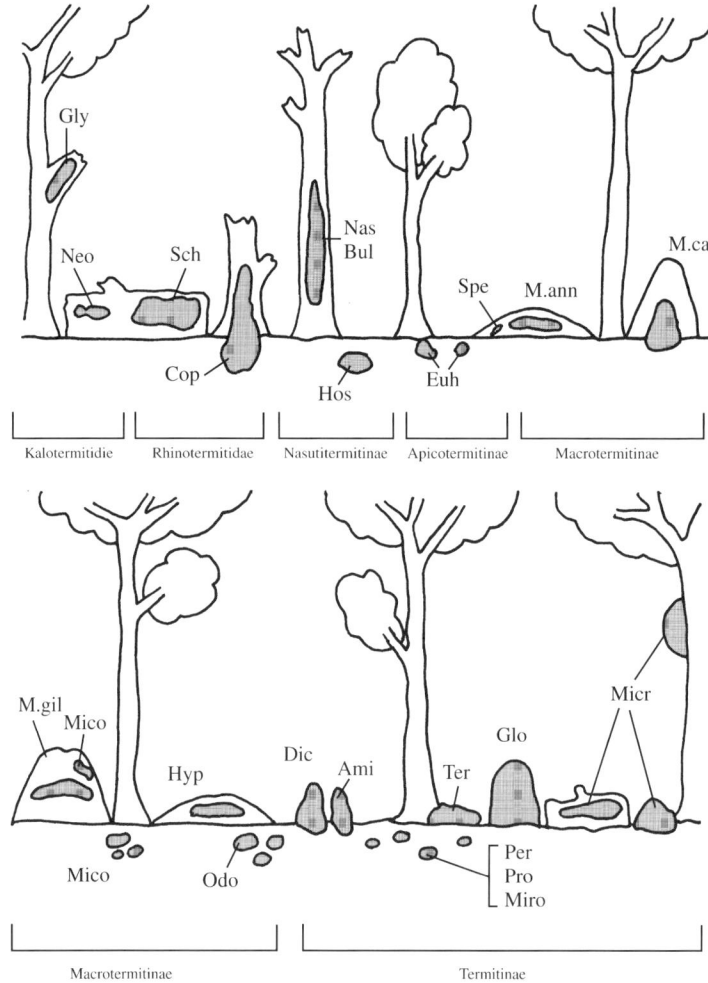

*Figure 2.25: The principal microhabitats of termites in a forest.* Abbreviations: *Gly*, Glyptotermes; *Neo*, Neotermes; *Sch*, Schedorhinotermes; *Cop*, Coptotermes; *Nas*, Nasutitermes; *Bul*, Bulbitermes; *Hos*, Hospitalitermes; *Spe*, Speculitermes; *Euh*, Euhamitermes; *M. ann*, Macrotermes annandalei; *M. car*, M. carbonarius; *M. gil*, M. gilvus; *Mico*, Microtermes; *Hyp*, Hypotermes; *Odo*, Odontotermes; *Ami*, Amitermes; *Dic*, Dicuspiditermes; *Per*, Pericapritermes; *Pro*, Procapritermes; *Miro*, Mirocapritermes; *Ter*, Termes; *Glo*, Globitermes; *Micr*, Microcerotermes.

*How to study termites*
More than 2 000 species of Isoptera have been recorded in the world (Eggleton 2000). The Order is divided into the 'lower' (Mastotermitidae, Kalotermitidae, Termopsidae, Hodotermitidae and Rhinotermitidae) and the 'higher' (Termitidae) group of families. Termites are social insects and make nests of various sizes. This makes their assessment difficult.

The principal microhabitats in which termites may be found in forests of a region are shown in Figure 2.25. They make their nests in the dead branches of living trees, on tree trunks, in standing dead trees, in fallen trees, on the ground surface and in the soil.

Termites may be collected with light traps (alates only), baits or by direct collection. The last of these methods is advocated here as the most straightforward and reliable. Eggleton and Bignell (1995) have examined the strengths and weaknesses of qualitative and quantitative methods for monitoring termite species diversity and propose a set of standard procedures. This protocol was described in detail by Jones and Eggleton (2000). This method was applied to forests in Borneo (Eggleton *et al.* 1997, 1999; Jones 1996, 2000) and Thailand (Davies 1997).

We propose a set of procedures to be used in IBOY-DIWPA based, in the first instance, on the proposals of Jones and Eggleton (2000) with additional measurements of biomass.

*Rapid assessment procedure: the belt transect method*
This segment of the assessment procedure requires 50 m tape (1), 100 m nylon string (preferably brightly colored), 2 m measuring pole (1), magnetic compass (1), coloured adhesive or flagging tape for marking off string sections, machetes (2), trowels (2), high-sided trays (2), fine forceps (2), stoppered vials approximately 1 cm × 5 cm (40+x), 80 % ethanol and labelling equipment

A transect should be laid out along a randomised compass bearing width using the measuring tape and pole. The transect is 100 m long and 2 m wide and is subdivided into 20 sections, each 5 m × 2 m marked by coloured tape. These should be clearly labelled at each end. Physical heterogeneity within the forest is a major determinant of overall termite diversity. Hence it is important that the transect includes, rather than avoids, any natural features along its track.

Sampling is conducted along a 100 m line sequentially. Each section is sampled by two trained people for 30 min (a total of 1 hour of collecting per section). In order to standardize sampling effort, the collectors should work steadily and continuously during each 30 min. collecting period. In each section the collectors search the following microhabitats, which are common sites for termites.

*Surface soil.* Twelve samples of surface soil (each about 12 × 12 cm, to 10 cm depth) should be scraped away. This should be examined in the sorting trays. Termites detected should be picked up by forceps and preserved. Accumulations of litter and humus at the base of trees and between buttress roots are also searched.

*Dead wood.* In each section of the transect a selection of the dead wood, varying from small twigs (ca. 5 mm diameter) to fallen logs, should be broken open and searched. Branches that are sound at one end may be termite-infested at the other. Small twigs should be broken open over a collecting tray to make termite collecting easier. Rotting wood partially incorporated into the topsoil will frequently contain termites.

*Tree trunks.* In each section of the transect all tree trunks should be inspected to a height of 2 m and searched for termite nests, runways or cartón sheeting. Runways and sheeting should be carefully scraped into a collecting tray and examined for insects.

*Termite nests.* Any termite mounds within each section of the transect should be broken open with minimal disturbance and termites within, collected. Many nests are host to termites other than the primary nest builder and so the periphery of the nest should also be examined. Subterranean nests are often found close to the soil surface or beneath fallen logs.

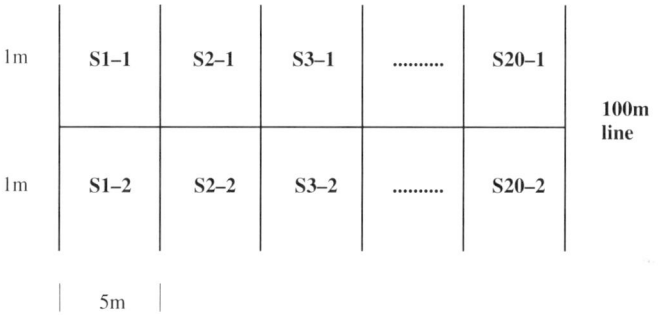

*Figure 2.26: The standardized protocol for measuring termite species diversity, the belt transect method (from Jones and Eggleton 2000).*

Collecting along a transect requires experience and the more experience a collector has the more complete the survey will be. New collectors reach the required level of competence by completing a training transect before they are used for the formal survey work.

Termites are eusocial insects and always co-occur with other members of their colony, often in very large numbers. When sampling, specimens from each termite population discovered must be collected. It is important to find individuals of the soldier castes as identification of most species is based on soldier morphology. The soldier/worker ratio is sometimes very low and it may not always be possible to locate soldiers. There are, indeed, some genera that are soldier-less. So workers should also be collected in all cases. Frequently these can be used for identification. More than one species will often occur in the same unit of microhabitat. Therefore, examples of all different morphotypes encountered must be collected. All termites collected should be placed in vials of 80 % ethanol. Each collector involved in the survey should carry a set of such vials.

To standardize sampling effort two trained collectors should spend thirty minutes in each section of the transect. A single transect will, accordingly, take ten hours to complete. A new collector should allow five hours on the training transect to develop the skills required to participate in the survey proper.

*Analysis.* Termites collected within the training transect cannot readily be used in any quantitative analyses but, clearly labelled, should be retained as they may assist in the later identification and related taxonomic processes. Termites from the survey transect should be sorted into recognizable taxonomic units (so-called 'RTU's). Sometimes direct recognition of the morphological features of named species will also be possible. Each 5 m section can be treated as a separate sample and species-accumulation curves constructed for the twenty units. For this purpose the following procedures are recommended:

- ten random sequences of sections are drawn from the transect without replacement;
- each of these random sequences is used to produce a separate species accumulation curve;
- the final accumulation curve is drawn using the means from each of the ten sequences;
- an estimate of the actual number of species in the area can be calculated using a first-order jack-knife estimator (see Magurran 1988; Colwell 1992).

*Quantitative sampling*
In contrast with the above procedures that are dedicated to the estimation of the species richness of termites in an area, we propose that an associated set of

surveys be used to estimate the termite biomass within the same area. This provides better information for understanding the relationship between species diversity and the ecosystem functions of the termites. This should be done using a set of soil cores along the belt transect together with a set of small quadrat surveys of both litter and soil within the transect area.

Fifty soil cores, each 10 cm × 10 cm × 30 cm deep should be taken at 2 m intervals along the 100 m transect. A further 20 quadrats, each 10 cm × 10 cm, are randomly located within the transect sections. At each of these quadrats the litter should first be collected. Then the soil from an adjacent area should be dug and a section 10 cm × 10 cm × 10 cm deep should be collected. This should be followed by a similar sample at depth 10–20 cm and then 20–30 cm. Any stones or wood encountered should be included in these samples. The termites in each of the soil cores and each of the segments of the transect surveys must be sorted from the soil and litter. The numbers and species involved will be relatively few and each sample should be able to be separated into morphospecies in this fashion. These samples should be preserved in total for later weight estimation. Estimates of total termite biomass per square metre of forest soil and litter can be derived by this method, stratified for depth and termite species.

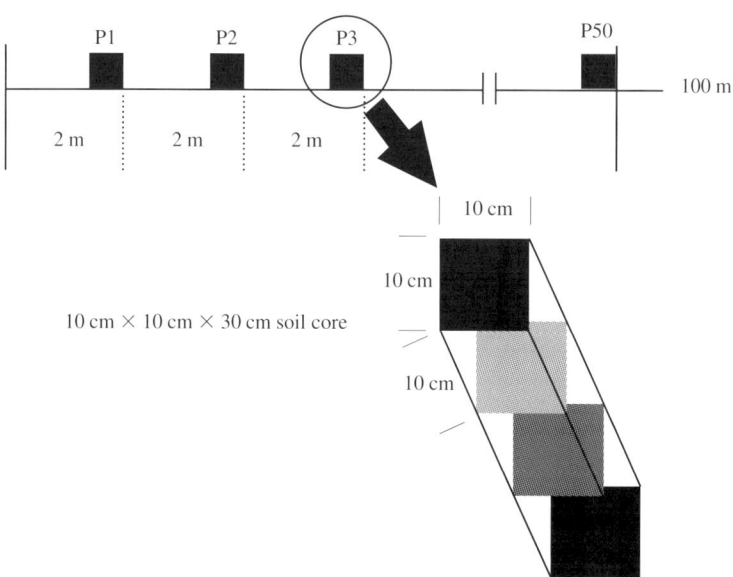

*Figure 2.27: The standardized protocol for evaluating termite biomass, the pit-digging method (based on Inone et al. 2001).*

*Table 2.9: Key works for termite identification in Asia.*

| Region | Reference | Coverage |
|---|---|---|
| Chine | Huang *et al.* (1989) | overview of national fauna |
| Thailand | Ahmad (1965) | overview of national fauna |
| Indomalayan | Ahmad (1958) | key to species |
| India | Roonwal & Chhotani (1989) | monograph of lower termites |
| Sabah | Thapa (1981) | overview of national fauna |
| Peninsular Malaysia | Tho (1992) | overview of national fauna |
| Burma | Krishna (1965) | overview of national fauna |

*Termite identification*
The taxonomy of termites is relatively advanced because of their economic importance. Some useful monographs exist; Table 2.9 summarizes some of the key works organized by region.

## 2.5.3 Handling, identification, storage and data management for arthropod surveys

Inventory is based on specimens. In principle then, for any inventory work conducted in accordance with these standards, all specimens should be stored for further study. This is particularly important where surveys have been incomplete or inexpert. Much new taxonomic material is present in samples and the preservation of this material is important for taxonomic purposes. Preserved specimens will also maintain genetic material and offer opportunities for the estimation of genetic biodiversity.

Biodiversity is the variety of life measured at the genetic, organism and ecological level. Accordingly there are many measures of biodiversity. Many desirable measures are difficult or impossible to obtain and in these cases we seek surrogates. Some surrogate measures will be more useful than others for solving particular questions. Despite limitations, species richness is probably the most widely used and acknowledged of surrogate measures. It integrates the three basic levels of diversity. It is relatively easy to measure and, despite some significant limitations, it has become a common currency in the study of biodiversity.

In complex forest ecosystems in particular, however, it can take a very long time to assess biodiversity to the species level following field surveys. There is general agreement that a large part of the world's species remain undescribed and that the coterie of taxonomists is too small to cope with the sheer volume of material in a 'traditional' fashion. These 'traditional methods' of occasional collecting expeditions, the gradual accumulation of material in museums and the subsequent production of monographs containing formal descriptions should, of course, continue but we need more rapid methods of assessment if the demands for information by governments and others for ecological 'answers' are to be met. This is especially the case for threatened ecosystems such as the many rainforests that are being cleared at such a rate that the opportunity to ever know much of their arthropod biodiversity is also threatened.

Accordingly we can separate our arthropod biodiversity assessment into three phases (Fig. 2.28). For the first phase samples will be sorted to the ordinal level made immediately after the sampling events. At the end of the IBOY-DIWPA a rapid evaluation at this level should be possible. A second phase will follow, in which some or all samples will be further sorted to the family or generic level. This second level can be made with the assistance of students, volunteers and other 'para-taxonomists'. This allows the fauna to be sorted into major guilds and the biodiversity can be evaluated in terms of its ecosystem-level functions. Finally the biodiversity assessment to the species level can be made in collaboration with groups of professional taxonomists. Ideally such groups would include those participating in taxonomically oriented biodiversity projects such as GaiaList 21 and Species 2000. All data produced at each stage should be publicly available. In addition the wide range of material collected from across the Asian/Australasian 'green belt' will attract the attention of taxonomists set on pursuing their own particular professional interests. They should of course be encouraged to work with IBOY-DIWPA samples so long as data and voucher specimens are returned to appropriate storage institutions in due course.

This section deals with the handling of specimens and data collected principally from the general trapping collections. Some information on specimen curation has been discussed earlier. The handling of the 'dry' light trap catches has been discussed as has the procedures for specialist ant collections. Accordingly this section refers to the other 'wet' material only.

Arthropod samples return from the field in a variety of conditions. Vials contain arthropods in ethanol, but they may also contain plant and soil remains, water and other contaminants. Sorting the arthropods from the rest can be a time-consuming process but the first task is to ensure that material remains well preserved until the sorting process can begin. If there is any possibility that the samples have been contaminated with water, they should be drained and filled with fresh 80 % ethanol. Similarly, any bottles that have dried up need to be replenished.

Equally important is that the presence and legibility of the data label for each unsorted sample must be checked. Again, if there is any doubt, new labels

# Forest Ecosystems

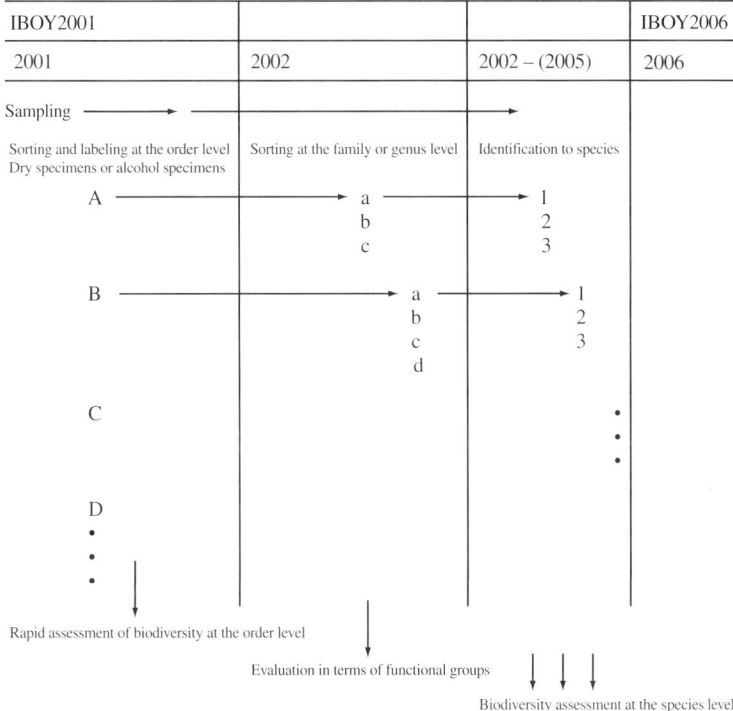

*Figure 2.28: Sample management and data analysis.*

should be attached while the provenance of the sample is still known. If the exact origin of the sample cannot be determined it must be discarded. Sorting samples without labels or with erroneous labels is simply a waste of time. On arrival in the field laboratory sample bottles should be 'signed in' in a notebook provided for the purpose that also records subsequent sorting. The logging of samples in and out enables any omissions and errors to be traced.

## Sorting and identification

Exact sorting procedures differ from worker to worker and each person sorting will develop his or her own detailed approach. In general a sample is emptied into a Petri dish which has a grid drawn or scratched onto its underside. The dish is then searched repeatedly under a binocular microscope and all arthropods are identified to Order in most cases. Once identified to this first level, arthropods are removed and placed in small vials with the members of each category for each sample being lumped together. It is most efficient to search the dish for one category such as the Diptera, or at most two, picking out and counting all the relevant individuals. Once the common groups have been removed a more general sorting for the rarer material can be made until no arthropods remain in the sample. In general fine watchmaker's forceps (#8) are used for picking up material from the dishes but occasionally a fine paintbrush may be substituted. Mites in particular often need a pipette for efficient handling.

As already noted, initial sorting should be to Order. We suggest two exceptions to this rule: ants may be separated from other hymenopterans because (a) it is easy to do and (b) different specialists generally deal with the two groups; Heteroptera and Homoptera also may be separated for similar reasons. Adult and immature animals belonging to the same group should be counted and stored separately for homometabolous insects but together for others.

It is important that each vial contain a clearly displayed label containing information on the site location, trap type and number, trapping date and taxon name. As these labels may need to last a substantial time until more detailed sorting can be carried out, and as they will be placed into a solution of 80% alcohol, it is preferable that they be laser printed, with, if necessary, any other details added in pencil or Indian ink by the sorter. The use of a laser printer also allows for the production of labels with barcodes. Each barcode represents a unique number containing the specimen information pertaining to the vial's contents. Using a barcode format allows easy tracking of specimens as the original sample is gradually broken up by sorting, dispatch to specialists, remounting and so forth.

Following the initial sorting, the whole set of vials from a particular sample are then bundled together using an elastic band and stored with the original sample label (when it survives). Material is usually returned from the field laboratory in this form. Once in its permanent home the bundles of vials are resorted so that all the vials containing insects of the same Order, or other category, from the same trapping method are put together.

The counts of each Order category of arthropods should be entered on a pre-printed tally sheet (Appendix) that also has the locality, date, type of trap and trap code entered onto it. Sorters tally individual organisms as they count them and then total each category. These tally sheets are VERY IMPORTANT. Once completed they are collected together and the information transferred to an Excel data file on a computer. The tally sheets should be retained and stored securely so that actual counts can be checked at a later date if any questions arise over accuracy.

## Storage

Surveys of this kind produce large quantities of material that soon present storage problems. The way in which this is handled is usually a compromise between the ideal and the possible. Ideally all vials should be placed in jars containing a base of cotton wool, kept part-filled with ethanol and should be stored at about 4°C until further sorting is required. In addition soft-bodied groups should be retained in 100 % not 80 % ethanol so that internal characters are retained for subsequent taxonomic analysis. In practice vials are often stored in jars of 80 % ethanol and most groups may be maintained adequately at room temperature in an air-conditioned laboratory. Collembola and Hymenoptera, where space permits, are stored in a freezer.

## Data storage and handling

The process of sorting several replicates of each trapping method to Order creates a great deal of data. A simple yet comprehensive and flexible approach to data storage and handling is essential.

One example of a database that can be used to keep track of the samples and the data associated with them is Filemaker™ Pro 4.1. This allows for a record per sample (at the Order level) with related databases for subgroups of specimens when sorted to lower taxonomic levels. It allows for tracking with a bottom-up (from individual sample to whole trap or site) or a top-down approach (from a whole trap or site to the individual samples it contained). There are a variety of packages available to fulfil this function.

Filemaker is particularly good because:
- it is inexpensive;
- it makes database creation simple;
- it makes data entry and report production easy;
- it operates on both Windows and Macintosh platforms. In fact, the same file can be accessed from either platform.

These points make it ideal for present purposes.

Microsoft Excel™ is used to tabulate the raw data. Excel is easy to use for both regular staff and volunteers in the field; it is readily available and widely used, allowing easy transfer of information to others; it facilitates easy data manipulation and simple statistical analysis; and, it allows for easy data transfer and reformatting for use in more sophisticated statistics and data-base packages.

An Excel work-book should be created for each trapping method with a separate work-sheet for each catch (e.g. a work-book for pitfall traps with a work-sheet for each day's catch). Each work-sheet has the site, time of year and name of trapping method at the top. Column 1 is a list of each order (or other taxa). It is important that the worksheet be created with a zero entry in every cell. These are then replaced with the counts of each category found for that replicate of the trapping method. If the zeros are omitted, Excel will calculate values such as means and standard errors incorrectly. Further work-sheets are then created with summary data such as totals, counts, means, standard errors and proportions. These can then be graphed or used for further analyses elsewhere.

If data entry is to be done alongside sorting in a field laboratory, standard Excel workbooks should be created before going into the field, with basic headings and zeros in place. In the field laboratory, provided conditions are suitable (i.e. there is electricity), a portable computer is ideal for the entry of data as sorting is completed. It is preferable that one person be responsible for this, to ensure consistency in all data entry.

### Backups

Another important aspect to data entry, one that cannot be over-emphasized, is backing up completed work (including work-in-progress). This can be achieved in two ways – first a print of completed data can be made and secondly a backup to some external medium, such as a floppy or zip disk, is made periodically as work progresses and again once counting is finished. Upon completion of data entry, it is prudent to make a separate backup to be stored separately off site. Should a major event, such as fire, happen in the normal work place, this raw data can then be re-accessed without difficulty.

## 2.6 Animal diversity: vertebrates

The Smithsonian Institution is in the process of producing handbooks on the assessment of biodiversity under the general editorship of Dr Mercedes Foster (Heyer *et al.* 1994; Wilson *et al.* 1996). The volumes that have appeared to date deal largely with the vertebrate groups (mammals and amphibians) and provide an important resource for designing vertebrate survey techniques for IBOY-DIWPA.

Vertebrate surveys differ fundamentally from those of invertebrates because they involve minimal collecting. Identification must be done in the field or later using photographs. For many groups pre-existing expertise is needed especially where surveys are to be done based on calls or scats rather than actual trapping. In addition it is important to note that many countries in the region have special regulations concerning the trapping and handling of vertebrates and many research institutions require that all work on vertebrates receive prior approval by their committees on animal experimentation. It is important both from a scientific and social point of view that these regulations and administrative procedures are adhered to by all workers involved in IBOY-DIWPA surveys. Visiting scientists should consult their local counterparts substantially ahead of their proposed in-country activities for information and assistance with local procedures. Particular care must be taken when working with rare or endangered taxa.

Another fundamental difference between vertebrate and invertebrate surveys is the existence of prior information about each region. For all of the major groups of terrestrial vertebrates (amphibians, reptiles, birds and mammals) indicative lists can be drawn up by judicious use of existing field-guides, handbooks and monographs for any country. A species list of mammals, birds, reptiles or amphibians may already exist for the region or an expert on a specific group may be available to draw up a preliminary list for the DIWPA site. This should be done before location specific surveys begin. Such indicative lists should include some indication of the relative abundance of the taxa concerned ("abundant", "common" or "rare"), some indication of conservation status ("threatened", "endangered", "at risk" etc.) and comment on the relative reliability of the information. Some taxa within each of the four major groups will have relatively poor information (e.g. smaller inconspicuous birds, small lizards, mice, rats and bats) and indicative lists will be less reliable in these cases.

On-site vertebrate surveys will involve up to four types of data collection:
   (i) The use of transect walks recording species by sight or call against a standard time and distance base. These require considerable prior expertise.
  (ii) The recording of the presence of species by their signs or artifacts. These may include the analysis of tracks, nests, scats, pellets or body remains.
 (iii) The use of specific traps to capture groups of animals. The preferred designs of traps differ from group to group but in each case a proper design must be established to ensure that the results obtained have quantitative validity.
 (iv) The collection of information by interview with local, often indigenous, peoples. This can be fraught with difficulties of communication about species, time scales or even intention. It can be used effectively for larger animals whose identity is not in doubt by either interviewer or interviewee.

The vertebrate survey should be undertaken in two steps:
1. Draw up a local indicative list from information available in print or from the knowledge of experts. The background information, such as the description of environmental variables or habitat, should also be provided (Table 2.10).
2. Undertake specific collecting or surveys (see below) as required.

For the description of wildlife habitats a simple 10-page form called "Identification and classification of wildlife habitats – Pro forma survey" is available in English, French or German. A more complex "Pro forma survey of rainforest vegetation" is 37 pages long and includes all significant structural components of tropical forest vegetation. Both are available on request from DIWPA Headquarters (Center for Ecological Research, Kyoto University).

## 2.6.1 Amphibians
**Pitfall and fence trapping**
Sampling using a combination of pitfall traps and drift fences (pitfall array) should be made 10 nights for each survey period. A minimum of ten arrays 10 m apart to a maximum of 30 arrays should be set in the field (Fig. 2.29). At core sites amphibians should be surveyed each time arthropod sampling is conducted, or at least once per quarter. All animals captured are euthanized and preserved under permits (special consideration will need to be given to species with special status, such as those listed by the Red-Data list as threatened or endangered). The humidity and air temperature should be recorded on each census occasion.

## 2.6.2 Reptiles
**Pitfall and fence trapping**
See the section for amphibians.

## 2.6.3 Birds
**Transect counts**
Transect counts along 1km of forest trail in an area of uniform habitat should be made in the early morning. The transect should be through the forest around the core plot of 1ha for IBOY. Counts should be made at least three times during the peak breeding season and repeated, if possible, in the non-breeding season. Weather conditions, time of day, speed of walking and methods of counting (using calls, sighting a set distance on either side of a trail, etc.) should be recorded. Distance to the bird is recorded, thus standard numbers versus distance decay curves can be plotted and used to estimate densities. Changes in the visibility of birds varying with species consciousness, light intensity, vegetation density and specific activity patterns will all affect the accuracy of such censuses. Ideally, counts should be made in calm weather, starting 30 minutes before sunrise for a duration of 2 hours. All birds seen or heard within a 50 m radius should be recorded on a map prepared beforehand to show the conspicuous features along the transect on a grid system. Depending on the birds' activity, it may be difficult to detect cryptic species while avoiding

*Table 2.10: Sample date form for a regional list of vertebrates.*

**DATE FORM for IBOY**

**Regional LIST for Vertebrates**

**SITE**  ( _____ )

**TARGET TAXA**  ❏ Mammals;  ❏ Birds;  ❏ Reptiles;  ❏ Amphibians

**GEOGRAPHIC**

COUNTRY  _____

STATE, DEPT, PROV, or ISLAND  _____

LOCALITY  _____ km  N, S, W, E of _____

LATITUDE _____  LONGITUDE _____  ALTITUDE _____

**HABITAT and ENVIRONMENT**

| | | | |
|---|---|---|---|
| GENERAL CLIMATE | ❏ wet; | ❏ humid; | ❏ arid; |
| | ❏ cyclic wet-dry | | |
| ENVIRONMENTAL ZONE | ❏ tropical; | ❏ subtropical; | ❏ warm-temperate; |
| | ❏ cool-temperate; | ❏ sub-boreal; | ❏ boreal; |
| | ❏ other _____ | | |
| GENERAL HABITAT | ❏ woodland; | ❏ open forest; | ❏ closed forest; |
| | ❏ cultivated; | ❏ shrubland; | ❏ grassland; |
| | ❏ savannah; | ❏ heath; | ❏ riparian; |
| | ❏ swamp; | ❏ marsh; | ❏ tundra; |
| | ❏ alpine; | ❏ taiga; | ❏ desert; |
| | ❏ dune; | ❏ beach; | ❏ semi-arid; |
| | ❏ shore; | ❏ urban; | ❏ suburban; |
| | ❏ island; | ❏ other _____ | |
| HABITAT TYPE | ❏ coniferous; | ❏ deciduous; | ❏ evergreen; |
| | ❏ secondgrow; | ❏ plantations; | ❏ scrub; |
| | ❏ thicket; | ❏ grove; | ❏ sandy; |
| | ❏ mangrove; | ❏ rocky; | ❏ cliff; |
| | ❏ bank; | ❏ cave; | ❏ other _____ |
| TEMPERATURE | ann. mean _____ | max. _____ | min. _____ |
| RAINFALL | ann. mean _____ | wet season _____ | dry season _____ |

**GENERAL DESCRIPTION**

SURVEY DURATION (d/m/y) _____ – _____

SURVEY SEASON  _____

SURVEY AREA  _____ ha

TECHNIQUES  ❏ direct obs. (transect count; others _____ );
 ❏ trapping (method: _____ );
 ❏ interviews;
 ❏ other _____

**RECORDIST**  ( _____ )

## Biodiversity Research Methods

**Trapping design**

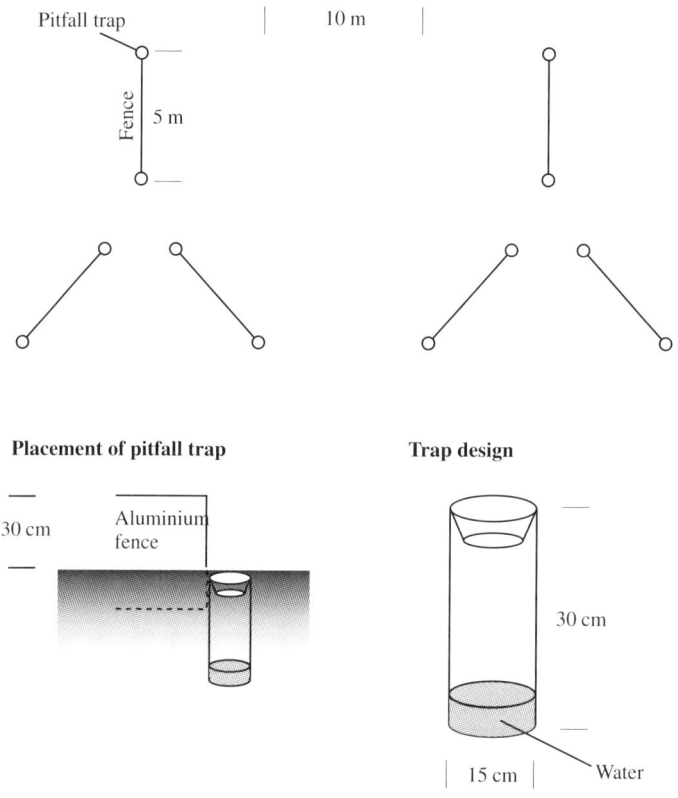

*Figure 2.29: Pitfall trapping for amphibian and reptile survey.*

duplicate entries of conspicuous ones. In some cases the point count may yield more consistent results than the transect count. It is advisable to consider the point count methods (described below) in combination with the transect count for accuracy and consistency before deciding on the most suitable method.

1. 5-minute point counts at 50 m interval in one direction followed by the same in the opposite direction along the transect.
2. 10-minute point counts at 50 m interval along the transect.
3. 20-minute point counts at 100 m interval along the transect.

For counts from a canopy walkway, point counts should be standardised at several vantage-points for 20 minutes in 5-minute intervals by scanning the foliage and canopy space. Record the direction and distance at which birds are sighted, together with the number of individuals and activities of the birds.

## 2.6.4 Mammals

The large diurnal mammals of a region are probably well known and lists can be prepared from local knowledge. For nocturnal mammals and species with cryptic habits the following procedures are suggested.

### Trapping for small mammals

Small mammal trapping should be conducted at all core and satellite sites. At each site trapping should be conducted around the time of arthropod sampling. However, it is recommend that the small mammal survey is done before the arthropod sampling, or at least two weeks afterwards, to avoid interference from disturbance by researchers. At core sites small mammals should be surveyed each time arthropod sampling is conducted, or at least once per quarter. It is recommended that site-organizers arrange for researchers with experience of small mammal trapping to conduct the surveys or at least for members to practice the techniques of setting the traps and handling the animals before the survey is conducted.

> RESEARCHERS MUST BE APPROPRIATELY VACCINATED AGAINST TETANUS, RABIES OR OTHER LOCAL DISEASES THAT CAN BE TRANSFERED THROUGH THE BITE OF A SMALL MAMMAL. RESEARCHERS ARE ALSO RESPONSIBLE FOR OBTAINING APPROPRIATE LICENCES OR PERMISSION FOR LIVE-TRAPPING OF SMALL MAMMALS.

### Traps

For the purposes of this section small mammals are considered to be those less than 2 kg and non-volant. Two types of trap should be used: (1) Wire-cage traps (approximately 20 cm × 20 cm × 30 cm) with a spring-loaded front-door can usually be obtained locally. The springs and wire must be strong enough to prevent the animals forcing their way out; (2, optional) 'Longworth' style box traps, with a tunnel entrance (approximately 4 cm × 4 cm entrance), can be obtained from equipment suppliers. Thirty traps of each type are required, including spares. If traps are new it is recommended that they be left out in the forest for 1–2 weeks. Gloves should be used when handling traps. This type of trap is effective and strongly recommended for the temperate and boreal sites.

## Grid lay-out

Trapping is done in the 1 ha plots. Starting at the 0,0 co-ordinate traps are laid out in a 5 × 5 grid at 20 m intervals using the plot markers. Place one of each type of trap within 3 m of the plot-marker and with a minimum of 2 m between them. Mark the trap with a bamboo pole or colored tape or both. After ground trapping is completed traps should be set off the ground from 3 m to canopy height (if possible) on approximately the same grid. Use aluminium ladders to get access and strap traps firmly to branches (strips of rubber inner-tube are useful for this).

## Bait

Wire-cage traps should be baited with slices of banana. Box traps should be baited with oats or barley and bedding material (hay or shredded cloth) should be placed in the box.

## Pre-baiting

Traps should be located and baited, but the trip-mechanisms not set (i.e., the doors should be secured open) for 2–3 days before trapping. During this time the traps should be checked once per day to replace bait.

## Trapping

Trapping is conducted over five consecutive days and nights. On the morning of the first day replace all bait and bedding and set the mechanisms. Traps should then be checked every evening and early morning, replacing bait and bedding as necessary. Always replace bait and bedding after a capture. Do not discard old bait and bedding in the plot. Put it in a plastic bag and dispose of it at the field centre. When ground trapping is complete replace all bait and bedding for the pre-baiting of traps in their off-ground positions.

## Handling of small mammals

Handling of small mammals is made easier if one person can hold the animal while another takes measurements and records data. The following data should be collected for all captures; date, trap type, plot co-ordinates, day of capture (1st–5th), height of trap, species identification, sex, sexual condition (males – swollen testes or not; female – vagina perforated or not, pregnant, or lactating) and weight (carry a 0–100 g and 0–2 kg balance). For species on which there are few studies available or those for which identification is doubtful, the following data should also be collected: number and position of teats if female and head and body, tail, hind-foot and ear (inside base to top) length (use calipers). Captures should not be released until they have been positively identified.

Mouse-sized mammals (< 100 g) are easily handled by emptying the trap into a clear-plastic sample-bag and closing the neck with one hand. The mammal can then be identified. The bag + mammal should be weighed with a spring-balance. The mammal can be held in a corner of the bag from outside while the other hand grabs the skin on the back of the neck between thumb and forefinger. Subtract the pre-weighed sample bag from the combined bag + mammal weight to determine the net weight of the mammal.

Larger rat sized mammals should be emptied into a strong clear-plastic bag and anaesthetized carefully with ether. Remove them as soon as they lose mobility. Handle them with thick-leather gloves.

Before release the animal should be individually marked so that recaptures can be recognized. We recommend using toenail clipping as it is easy to make a simple code for individual marking and the clipping does not injure the animal.

Only individuals that cannot be identified and are believed to be possible new species should be collected.

## Data handling and storage

Initially a species-list with an indication of the status of each species (e.g. uncertain, endangered, rare, common, abundant) should be prepared and forwarded to the DIWPA headquarters (Center for Ecological Research, Kyoto University). A proposal for further study for each group of vertebrates should be made, based on preliminary work at the site.

The protocol for data storage and the procedure for more detailed surveys will be determined following the initial data collection.

## Identification of vertebrates

The names of the experts responsible for the identification of species should accompany the species-lists. If trained personnel draw up the species-lists, specify the keys or guidebooks used for identification. Table 2.11 summarizes some key works on each of the four vertebrate groups organized by regions.

# Biodiversity Research Methods

*Table 2.11: List of identification manuals for Vertebrates.*

### Amphibians and Reptiles
Alcala, A.C. (1986) *Guide to Philippine Flora and Fauna, 10, Amphibians and Reptiles*. NRMC, Philippines.
Barker, J., Grigg, G.C. and Tyler, M.J. (1995) *A Field Guide to Australian Frogs* (2nd edn.). Surrey Beatty and Sons., Australia.
Capula, M. (1989) *Guide to the Reptiles and Amphibians of the World*. (Behler, J.L. ed) Simon & Schuster, USA
Cogger, H.G. (1996) *Reptiles and Amphibians of Australia*. Reed New Holland, Australia.
Inger, R.F. and Lian, T.F. (1996) *The Natural History of Amphibians and Reptiles in Sabah*. Natural History Publications, Borneo, Malaysia.
Inger, R.F. and Stuebing, R. (1997) *A Field Guide to the Frogs of Borneo*. Natural History Publications, Borneo, Malaysia.
Lim, K.K.P. (1992) *A Guide to the Amphibians and Reptiles of Singapore*. Singapore SC Guides, Singapore
Menzies, J.I. (1976) *Handbook of Common New Guinea Frogs*. Wau Ecology Institute, Papua New Guinea.
Stuebing, R. and Inger, R.F. (1999) *A Field Guide to the Snakes of Borneo*. Natural History Publications, Borneo, Malaysia.
Tiwari, S.K. (1991) *Zoogeography of Indian Amphibians, Distribution, Diversity and Spatial Relationship India*, India.

### Birds
Beehler, B.M., Pratt, T.K. and Zimmerman, D.A. (1986) *Birds of New Guinea*. Princeton University Press, USA.
Bhushan, B., Fry, G., Hibi, A., Mundkur, T., Prawiradilaga, D.M., Sonobe, K. and Usui, S. (1993) *A Field Guide to the Waterbirds of Asia*. Wild Bird Society of Japan and the Asian Wetland Bureau, Japan.
Jeyarajasingam, A. and Pearson, A. (1999) *A Field Guide to the Birds of West Malaysia and Singapore*. Oxford University Press, USA.
King, B., Woodcock, M. and Dickinson, E.C. (1998) *Collins Field Guide to the Birds of South East Asia*. Harper Collins, UK.
Kingsford, R. (1991) *Australian Waterbirds: A Field Guide*. Kangaroo Press, Australia.
MacKinnon, J. (1988) *Field Guide to the Birds of Java and Bali*. Gadjah Mada UP, Indonesia.
MacKinnon, J. (1993) *A Field Guide to the Birds of Borneo, Sumatra, Java and Bali*. Oxford University Press, USA.
Simpson, K. and Day, N. (1999) *Field Guide to the Birds of Australia* (6th ed.). Penguin Books, Australia.
Wild Bird Society of Japan (1982) *A Field Guide to the Birds of Japan*. Wild Bird Society of Japan, Japan.
Won Pyong-Oh (1996) *A Field Guide to the Birds of Korea*, South Korea.
Xu Wei-shu, Zhao Zheng-kai, Zheng Guang-mei, Yan Chang-wei and Tan Yao-kuang (1996) *A Field Guide to the Birds of China*. Kingfisher Press, Taiwan.

### Mammals
Corbet, G.B. and Hill, J.E. (1992) *The Mammals of the Indomalayan Region: a Systematic Review*. Oxford University Press, USA.
Davies, G. and Payne, J. (1982) *A Faunal Survey of Sabah*. WWF, Malaysia.

*Table 2.11: (continued)*

**Mammals (continued)**
Jones, G.S. and Jones, D.B. (1976) *A Bibliography of the Land Mammals of Southeast Asia.* Bishop Museum, USA.
Lee, S.S. (1995) *A Guidebook to Pasoh.* Forest Research Institute, Malaysia.
Lekagul, B. and McNeely, J.A. (1988) *Mammals of Thailand* (2nd edn.). Darnsutha Press, Thailand.
Medway, L. (1983) *The Wild Mammals of Malaya (Peninsular Malaysia) and Singapore* (2nd edn., reprinted with corrections). Oxford University Press, Malaysia.
Mohd. Momin Khan (1992) *Mamalia Semenanjung Malaysia.* Wildlife Department of Malaysia, Malaysia.
Natural Learning Pty Ltd (1999) *Mammals of Australia.* Natural Learning, Australia.
Payne, J. and Andau, M. (1991) *Large Mammals in Sabah. In* Kiew, R. (ed) *The State of Nature Conservation in Malaysia.* Malayan Nature Society, Malaysia.
Payne, J., Francis, C.M. and Phillipps, K. (1985) *A Field Guide to the Mammals of Borneo.* The Sabah Society, Malaysia.
Wee, Y.C. and Ng, P.K.L. (1994) *A First Look at Biodiversity in Singapore.* National Council on the Environment, Singapore.
Wilson, D.E. and Reeder, D.M. (1993) *Mammal Species of the World.* Smithsonian Institution Press, USA.

# References

Abe, T. 1979 Studies on the distribution and ecological role of termites in a lowland rain forest of West Malaysia. 2. Food and feeding habits of termites in Pasoh Forest Reserve. *Japanese Journal of Ecology* 29: 121–35.

Abe, T. & Matsumoto, T. 1979 Studies on the distribution and ecological role of termites in a lowland rain forest of West Malaysia. 3. Distribution and abundance of termites in Pasoh Forest Reserve. *Japanese Journal of Ecology* 29: 337–51.

Abe, T. 1984 Colonization of the Krakatau Islands by termites (Insecta: Isoptera). *Physiology and Ecology Japan* 21: 63–88.

Abe, T. 1987 Evolution of life types in termites. In: Kawano, S., Connell, J.J.H. & Hidaka, T. (eds), *Evolution and Coadaptation in Biotic Communities.* University of Tokyo Press, Tokyo, pp. 125–48.

Ahmad, M. 1958 Key to the Indomalayan termites. *Biologia* 4: 33–198.

Ahmad, M. 1965 Termites (Isoptera) of Thailand. *Bulletin of the American Museum of Natural History* 131: 3–113.

Akhtar, M.S. 1975 Taxonomy and zoogeography of the termites (Isoptera) of Bangladesh. *Bulletin of the Department of Zoology University of Panjab (New Series)* 7: 1–199.

Bogush, P.P. 1958 Some results of collecting click beetles (Coleoptera, Elateridae) with light traps in Central Asia. *Entomological Review* 39: 291–9.

Bolton, B. 1994 *Identification Guide to the Ant Genera of the World.* Harvard University Press, Cambridge.

Bowden, J. 1973 The influence of moonlight on catches of insects in light traps. Part I The moon and moonlight. *Bulletin of Entomological Research* 63: 113–28.

Bowden, J. & Church, B.M. 1973 The influence of moonlight on catches of insects in light traps. Part II: The effect of moon phase on light trap catches. *Bulletin of Entomological Research* 63: 129–42.

Briggs, J.B. 1961 A comparison of pitfall trapping and soil sampling in assessing populations of two species of ground beetles (Col.: Carabidae). *Report of the East Malling Research Station* 1960: 108–12.

Brock, J.P. 1971 A contribution towards an understanding of the morphology and phylogeny of the Ditrysian Lepidoptera. *Journal of Natural History* 5: 29–102.

Brown, J.H. 1995 *Macroecology*. University of Chicago Press, Chicago.

Chhotani, O.B. 1970 Taxonomy, zoogeography and phylogeny of the genus *Cryptotermes* (Isoptera: Kalotermitidae) from the Oriental Region. *Memoirs of Zoological Survey of India* 15: 1–81.

Collins, N.M. 1984 The termites (Isoptera) of the Gunung Mulu National Park, with a key to the genera known from Sarawak. *The Sarawak Museum Journal* 30: 65–87.

Colwell, R.K. 1992 Human aspects of biodiversity: an evolutionary perspective. International Union of Biological Science, Monograph No.8. In: Solbrig, O.T., van Emden, H.M. & van Oordt, P.G.W.J. (eds.), *Biological Diversity and Global Change*. IUBS Press, Paris, pp. 209–22.

Common, I.F.B. 1990 *The Moths of Australia.* University of Melbourne Press, Melbourne.

Davies, R.G. 1997 Termite species richness in fire-prone and fire-protected dry deciduous dipterocarp forest in Doi Suthep-Pui National Park, Northern Thailand. *Journal of Tropical Ecology* 13: 153–60.

Davies, S.J.J.F. 1984 *Methods of Censusing Birds in Australia*. R.A.O.U. Report No. 7.

Eggleton, P. & Bignell, D.E. 1995 Monitoring the response of tropical insects to changes in the environment: troubles with termites. In: Harrington, R. & Stork, N.E. (eds.), *Insects in a changing environment*. Academic Press, London, pp. 473–97.

Eggleton, P., Bignell, D.E., Sands, W.A., Waite, B., Wood, T.G. & Lawton, J.H. 1995 The species richness of termites (Isoptera) under differing levels of forest disturbance in the Mbalmayo Forest Reserve, Southern Cameroon. *Journal of Tropical Ecology* 11: 85–98.

Eggleton, P., Homathevi, R., Jeeva, D., Jones, D.T., Davies, R.G. & Maryati, M. 1997 The species richness and composition of termites (Isoptera) in primary and regenerating lowland deiperocarp forest in Sabah, East Malaysia. *Ecotropica* 3: 119–28.

Eggleton, P., Homathevi, R., Jones, D.T., MacDonald, J.A., Jeeva, D., Bignell, D.E., Davies, R.G. & Maryati, M. 1999 Termite assemblages, forest disturbance and greenhouse gas fluxes in Sabah, East Malaysia. *Philosophical Transactions of the Royal Society of London, Series B* 354: 1791–1802.

Eggleton, P. 2000 Global patterns of termite diversity. In: Abe, T., Bignell, D.E. & Higashi, M. (eds.), *Termites: Their Symbiosis, Sociality and Global Diversification.* Kluwer Academic Publishers, Dordrecht, pp. 25–51.

Erwin, T.L. 1982 Tropical forests: their richness in Coleoptera and other arthropod species. *Coleopterists Bulletin* 36: 74–5.

Erwin, T.L. 1990 Canopy arthropod biodiversity: a chronology of sampling techniques and results. *Revista Peruana de Entomologia* 32: 71–7.

Ford, J. 1937 Fluctuations in natural populations of Collembola and Acarina. *Journal of Animal Ecology* 6: 98–111.

Frost, S.W. 1952 Light traps for insect collection, survey and control. *Bulletin of the Pennsylvania Agricultural Experimental Station* 550: 1–32.

Frost, S.W. 1957 The Pennsylvania light trap. *Journal of Economic Entomology* 50: 287–92.

Gagné, W.C. 1979 Canopy-associated arthropods in *Acacia kea* and *Metrosideros* tree communities along an altitudinal transect on Hawaii Island. *Pacific Insects* 21: 56–82.

Geier, P.W. 1960 Physiological age of codling moth females (*Cydia pomonella*) caught in bait and light traps. *Nature (London)* 265: 415–21.

Greenslade, P.J.M. 1973 Sampling ants with pitfall traps: digging-in effects. *Insectes Sociaux* 20: 343–53.

Greenslade, P. and Greenslade, P.J.M. 1971 The use of baits and preservatives in pitfall traps. *Journal of the Australian Entomological Society* 10: 253–60.

Gressitt, J.L. and Gressitt, M.K. 1962 An improved Malaise trap. *Pacific Insects* 3: 549–55.

Heyer, R.W., Donnely, M.A., McDiarmid, R.W., Hayek, L-A.C. & Mercedes, S.F. 1994 *Measuring and Monitoring Biological Diversity: Standard*

*Methods for Amphibians.* BIODIVERSITY HANDBOOK SERIES 1, Smithsonian Institutional Press, USA.

Hill, C.J. 1999 Butterflies in communities. In: Kitching, R.L., Scheermeyer, E., Pierce, N.E. & Jones, R.E. (eds.), *The Biology of Australian Butterflies.* CSIRO, Melbourne.

Hollingsworth, J.P., Briggs, C.P., Glick, P.A. & Graham, H.M. 1961 Some factors influencing light trap collections. *Journal of Economic Entomology* 54: 305–8.

Hölldobler, X. & Wilson, E.O. 1990 *The Ants.* The Belknap Press of Harvard University Press, Cambridge, Mass.

Huang, F.S., Li, G.X. & Zhu, S.M. 1989 *The Taxonomy and Biology of Chinese Termites – Isoptera.* Tianze Press, Beijing.

Ikehara, S. 1966 Distribution of termites in the Ryukyu Archipelago. *Bulletin of Arts & Science Division, University of the Ryukyus, Mathematics & Natural Sciences* 9, 49–178.

Inoue, T., Takematsu, Y., Hyodo, F., Sugimoto, A., Yamada, A., Klangkaew, C., Kirtibutr, N. and Abe, T. 2001 The abundance and biomass of subterranean termites in a dry evergreen forest of Northeast Thailand. *Sociobiology* 37: 41–52.

Jones, D.T. 1996 A quantitative survey of the termite assemblage and its consumption of food in lowland mixed dipterocarp forest of Brunei Darussalam. In: Edwards, D. S. (ed.), *Tropical Rainforest Research – Current Issues.* Kluwer Academic Publishers, Dordrecht, pp. 297–305.

Jones, D.T. 2000 Termite assemblages in two distinct montane forest types at 1000m elevation in the Maliau Basin, Sabah. *Journal of Tropical Ecology* 16: 271–86.

Jones, D.T. & Eggleton, P. 2000 Sampling termite assemblages in tropical forests: testing a rapid biodiversity assessment protocol. *Journal of Applied Entomology* 37: 191–203.

Juillet, J.A. 1963 A comparison of four types of trap used for capturing flying insects. *Canadian Journal of Zoology* 41: 219–23.

Kitching, R.L., Bergelson, J.M., Lowman, M.D., McIntyre, S. & Carruthers, G. 1993 The biodiversity of arthropods from Australian rainforest canopies: general introduction, methods, sites and ordinal results. *Australian Journal of Ecology* 18: 181–91.

Krishna, K. 1965 Termites (Isoptera) of Burma. *Am. Mus. Novitates* 2210: 1–34.

Lee, E.E. & Wood, T.G. 1971 *Termites and Soils.* Academic Press, London.

Luff, M.L. 1975 Some factors affecting the efficiency of pitfall traps. *Oecologia (Berlin)* 19: 345–57.

Macfadyen, A. 1955 A comparison of methods for extracting soil arthropods. In: Kevan, D.K.McE. (ed.), *Soil Zoology*. University of Nottingham, Nottingham.

Macfayden, A. 1962 Soil arthropod sampling. In: Cragg, J.B. (ed.), *Advances in Ecological Research* 1. Academic Press, London.

Magurran, A.E. 1988 *Ecological Diversity and its Measurement*. Princeton University Press, Princeton.

Malaise, R. 1937 A new insect trap. *Entomologische Tidjschrift* 58: 148–60.

Martin, J.L. 1966 The insect ecology of red pine plantations in central Ontario. IV. The crown fauna. *Canadian Entomologist* 98: 10–27.

Maryati, M. 1994 Can ant be a bioindicator? *Proceedings of the 3rd Symposium of Applied Biology*, pp. 128–9.

Matsumoto, T. and Abe, T. 1979 The role of termites in an equatorial rain forest ecosystem of West Malaysia II. Leaf litter consumption on the forest floor. *Oecologia* 38: 261–74.

Minet, J. 1991 Tentative reconstruction of the ditrysian phylogeny (Lepidoptera: Glossata). *Entomologica Scandinavica* 2: 69–95.

Mitchell, B. 1963 Ecology of two carabid beetles, *Bembidion lambros* (Herbst.) and *Trechus quadristriatus* (Schrank). II. *Journal of Animal Ecology* 32: 377–92.

Moore, N.W. 1975 Butterfly transects in a linear habitat. *Entomologists' Gazette* 26: 71–8.

Morimoto, K. 1973 Termites from Thailand. *Bulletin of the Governmental Forest Experiment Station* 257: 57–80.

Mueller-Dombois, D. 1998 Vegetation and ecosystem research for biodiversity conservation in the Pacific islands. *Pacific Science Association Information Bullletin* 50: 1–10.

Neilsen, E.S., Edwards, E.D. & Rangsi, T.V. 1996 *Checklist of the Lepidoptera of Australia*. CSIRO, Melbourne.

Orr, A.G. & Haeuser, C.L. 1996 Temporal and spatial patterns of butterfly diversity in a lowland tropical rainforest. In: Edwards, D.S., Booth W.E. & Choy, S.C. (eds.), *Tropical Rainforest Research – Current Issues*. Kluwer Academic Publishers, Dordrecht, pp. 125–38.

Paris, O.A. & Pitelka, F.A. 1962 Population characteristics of the terrestrial isopod *Armadillidium vulgare* in California grassland. *Ecology* 43: 229–48.

Pollard, E., Elias, D.O., Skelton, M.J. & Thomas, J.A. 1975 A method of assessing the numbers of butterflies in Monk's Wood National Nature Reserve. *Entomologists' Gazette* 26: 79–88.

Roberts, R.H. 1970 Color of Malaise traps and collection of Tabanidae. *Mosquito News* 39: 567–71.

Roberts, H.R. 1973 Arboreal Orthoptera in the rain forests of Costa Rica collected with insecticide: A report on grasshoppers (Acrididae) including new species. *Proceedings of the Academy of Natural Sciences, Philadelphia* 125: 46–66.

Robinson, H.S. & Robinson, P.J.M. 1950 Some notes on the observed behaviour of Lepidoptera in flight in the vicinity of light-sources together with a description of a light-trap designed to take entomological samples. *Entomologists' Gazette* 1: 3–15.

Roonwal, M.L. 1970 Termites of the oriental region. In: Krishna, K. & Weesner, F. M. (eds.), *Biology of Termites*. Academic Press, New York, pp. 315–91.

Roonwal, M.L. and Chhotani, O.B. 1989 *The Fauna of India and Adjacent Countries. Isoptera (Termites) Vol. 1*. Zoological Survey of India, Calcutta.

Roonwal, M.L. & Maiti, P.K. 1966 Termites from Indonesia, including West Irian. *Treubia* 27: 63–140.

Southwood, T.R.E., Moran, V.C. & Kennedy, C.E.J. 1982a The richness, abundance and biomass of the arthropod communities of trees. *Journal of Animal Ecology* 51: 635–50.

Southwood, T.R.E., Moran, V.C. & Kennedy, C.E.J. 1982b The assessment of arboreal insect fauna: comparisons of knockdown sampling and faunal lists. *Ecological Entomology* 7: 331–40.

Stork, N. E. 1987a Guild structure of arthropods from Bornean rain forest trees. *Ecological Entomology* 12: 69–80.

Stork, N. E. 1987b Arthropod faunal similarity of Bornean rain forest trees. *Ecological Entomology* 12: 219–26.

Stork, N.E. 1988 Insect diversity: facts, fiction and speculation. *Biological Journal of the Linnean Society* 35: 321–37.

Tan, M.W., Kirton, L.G. and Kirton, C.G. 1992 Composition and distribution of butterflies along Sungei Kinchin and its vicinity. In: Yeong, Y.S. & Win, L.S. (eds.), *In Harmony with Nature*. Malayan Nature Society, Kuala Lumpur, pp. 193–212.

Taylor, L.R. & Taylor, R. 1977 Aggregation, migration and population mechanisms. *Nature (London)* 265: 415–21.

Thapa, R.S. 1981 Termites of Sabah. *Sabah Forest Record* 12: 1–374.

Tho, Y.P. 1992 Termites of Peninsular Malaysia. *Malaysian Forest Records* 36: 1–224.

Toda, M.J. 1977 Two new "retainer" bait traps. *Drosophila Information Service* 52: 180.

Townes, H. 1962 Design for a Malaise trap. *Proceedings of the Entomological Society of Washington* 64: 253–62.

Watanabe, H., Takeda, H. and Ruaysoongnern, S. 1984 Termites of northeastern Thailand with special reference to changes in species composition due to shifting cultivation. *Memoirs of the College of Agriculture, Kyoto University* 125: 45–57.

Wilson, D.E., Cole, F.R., Nichols, J.D., Rudran, R. & Foster, M.S. 1996 *Measuring and Monitoring Biological Diversity: Standard Methods for Mammals.* BIODIVERSITY HANDBOOK SERIES 2, Smithsonian Institution Press, USA.

Wood, T.G. & Sands, W.A. 1978 The role of termites in ecosystems. In: Brian, M.V. (ed.), *Production Ecology of Ants and Termites.* Cambridge University Press, Cambridge, pp. 245–92.

Yamane, Sk., Itino, T. & Rahman, N. 1996 Ground ant fauna in a Bornean dipterocarp forest. *Raffles Bulletin of Zoology* 44: 253–62.

# Biodiversity Research Methods

# Appendix: Sample Arthropod Tally Sheet

Note: this tally sheet includes only those Orders we have encountered to date in our surveys. Of course additional taxa may be encountered from time to time.

Site location e.g. Tomakomai Experimental Forest 1 ha
Sample type e.g. Light trap 1-CANOPY
This is the date of collection NOT the date sorted
Person who does the sorting

LOCATION: _____
SAMPLE: _____
DATE ON LABEL: _____
SORTED BY: _____

| ORDER | TALLY | TOTAL |
|---|---|---|
| Collembola | | |
| Diplura | | |
| Archaeognatha | | |
| Thysanura | | |
| Ephemeroptera | | |
| Odonata | | |
| Plecoptera | | |
| Blattodea | | |
| Isoptera | | |
| Mantodea | | |
| Orthoptera | | |
| Dermaptera | | |
| Phasmatodea | | |
| Embioptera | | |
| Psocoptera | | |
| Homoptera | | |
| Heteroptera | | |
| Thysanoptera | | |
| Neuroptera | | |
| Coleoptera | | |
| Diptera | | |
| Lepidoptera | | |
| Thrichoptera | | |
| Ants | | |
| Other Hymenoptera | | |
| Isopoda | | |
| Amphipoda | | |
| Araneida | | |
| Acari | | |
| Opiliones | | |
| Pseudoscorpiones | | |
| Chilopoda | | |
| Diplopoda | | |
| Symphyla | | |

# Drafting team

T. Abe, Center for Ecological Research, Kyoto
Y. Hashimoto, Museum of Nature and Human Activities, Kobe
Y. Hirai, Hokkaido University, Sapporo
K. Hurley, Griffith University, Brisbane
N. Inari, Hokkaido University, Sapporo
T. Itioka, Nagoya University, Nagoya
J. Kikkawa, Queensland University, Brisbane
R.L. Kitching, Griffith University, Brisbane
M. Laidlaw, Griffith University, Brisbane
M. Murakami, Hokkaido University, Sapporo
H. Takeda, Kyoto University, Kyoto
Y. Takematsu, Kyoto Institute of Technology, Kyoto
S. Tanabe, Hokkaido University, Sapporo
M.J. Toda, Hokkaido University, Sapporo
I. Turner, Botanic Gardens, Singapore
G. Vickerman, Griffith University, Brisbane
Sk. Yamane, Kagoshima University, Kagoshima
T. Yoshida, Hokkaido University, Sapporo

# Authors of sections

| | |
|---|---|
| Introduction, Site selection | R. Kitching, T. Itioka, M.J. Toda, T. Yoshida, I. Turner and M. Laidlaw |
| Environmental variables | T. Yoshida and I. Turner |
| Plants | T. Yoshida and I. Turner |
| Animals: Arthropods: | |
|     Light traps | K. Hurley, R.L. Kitching and Y. Hirai |
|     Malaise traps | M.J. Laidlaw and R.L. Kitching |
|     Window traps | N. Inari and M.J. Toda |
|     Canopy knockdown | R.L. Kitching and K. Hurley |
|     Bark spraying | G. Vickerman and R.L. Kitching |
|     Pitfall traps | G. Vickerman and R.L. Kitching |
|     Litter sampling | R.L. Kitching and K. Hurley |
|     Soil animals | H. Takeda and G. Takaku |
| Animals: Arthropods: Surveys of selected taxa | |
|     Drosophilidae | S. Tanabe and M.J. Toda |

| | |
|---|---|
| Lepidoptera | R.L. Kitching |
| Spiders | R.L. Kitching (based on Coddington *et al.* 1991) |
| Ants | S. Yamane and Y. Hashimoto |
| Termites | T. Abe and Y. Takematsu |
| Handling and identification of arthropod specimens | |
| | R.L. Kitching, T. Itioka and M.J. Toda |
| Samples and data management | K. Hurley and R.L. Kitching |
| Animals: Vertebrates | J. Kikkawa, M. Murakami and R. Harrison |
| Appendix | K. Hurley and Y. Hirai |

# Chapter 3: Freshwater Ecosystems

*Chapter Editors: Katsuki Nakai, Oleg A. Timoshkin, Dede I. Hartoto, Sulastri, Atsushi Doi, Toshio Iwakuma, Natalia G. Melnik, Masahide Yuma & Masami Nakanishi*

## 3.1 General Strategy

### 3.1.1 Endangered freshwater ecosystems

With rapid and drastic environmental changes occurring on both global and local scales, the concept of "biodiversity", or biological diversity, has recently attracted great attention, not only academically but also publicly and politically. In particular, freshwater ecosystems (lakes and rivers) and their inhabitants are regarded as being quite vulnerable and fragile under various anthropogenic impacts: the water in freshwater ecosystems, being the vital medium shared among inhabitants, is easily polluted both trophically and toxically due to the relatively small volume of water mass and the relatively large influence of human intervention. The shores and bottoms of lakes are often modified or destroyed through reclamation, embankment, and dredging. The shores and bottoms of rivers are, often more intensively, modified to artificially enhance the efficiency of water flow by making the river flow straighter. The construction of dams and weirs to regulate river flows and water levels has resulted in significant habitat modification. Maintenance and preservation/conservation of the biodiversity of freshwater ecosystems in lakes and rivers are thus becoming issues of high priority.

In terms of spatial isolation, lakes and rivers for aquatic organisms are analogous to islands for terrestrial organisms. Biogeographically, more than a few lakes and river systems are inhabited by endemic species that are unique to the respective water systems. Typically, the so-called "ancient lakes" – those that have existed for an exceptionally long time – provide exemplary cases of markedly high endemicity. As mentioned above, however, many freshwater environments and their biodiversity have been subject to serious threat of habitat deterioration due to various anthropogenic impacts, including:

1) eutrophication: caused by excessive nutrient (nitrogen, phosphorus, etc.) loading, deriving from soil erosion and surplus usage of nutrients in agriculture and forestry activities and from sewage water,

2a) organic pollution: caused by the noxious materials supplied through inflows or aerosol,
2b) chemical pollution: caused by inflow chemicals such as pesticides from farmlands, chemical wastes from factories and households, and heavy metals from mineries (sometimes causing serious chemical accumulation within the body of animals at higher trophic levels through food chains),
3) siltation or sedimentation: caused by suspended matter in inflow water (for which erosion in the watershed area is chiefly responsible),
4) modification or destruction of physical habitats: including reclamation, embankment, dredging, dam-construction, etc.,
5) excessive exploitation (overfishing) or aquaculture (including artificial stocking): targeting valuable species in commercial fisheries,
6) invasion and establishment of non-indigenous species: through accidental intrusions or intentional introductions (including domestic translocations), which sometimes lead to a depletion of original biodiversity sufficient to be regarded as a form of "bio-hazard",
7) artificial control of lake water levels and river water flows: conducted for acquisition of water resources or prevention of flooding, often providing unnatural environmental fluctuation patterns,
8) various effects from global climate changes (typically reflected in local temperature regimes and precipitation patterns).

These various human-induced impacts affect freshwater ecosystems not singly but in compound, making it almost impossible to evaluate the (even relative) contribution of a single factor to the deterioration of the ecosystems.

## 3.1.2 Biodiversity investigation in freshwater ecosystems

In spite of the extent of these endangered situations, biodiversity in freshwater ecosystems has been rather poorly investigated. Our knowledge and information about biodiversity inventories in the DIWPA region are limited partly due to the paucity or poor standardization of methodology.

For some lakes, accumulated data sets on planktonic communities in the offshore/pelagic zone and benthic communities on the deep (soft) bottom zone are available, these having been surveyed in conventional (rather standardized) methods, chiefly utilizing on-board samplers. Even for well-investigated lakes, though, little is known about the biodiversity of the littoral zone – believed to be the most species-rich and greatly diversified, owing to high primary productivity and extensive topographic heterogeneity – mostly because of practical difficulties in collecting samples, especially quantitative samples.

Recently, owing to technical and instrumental advance in SCUBA (Self-Contained Underwater Breathing Apparatus) diving, these practical difficulties can be largely resolved through direct underwater observation and sampling in the littoral zone, provided that the water is sufficiently transparent and overall safety is ensured.

In addition to direct investigation, the use of remote-controlled video devices and onboard echosounders provide useful supplementary information for describing an underwater landscape.

## 3.2 Unique framework for the IBOY-DIWPA

### 3.2.1 Freshwater environments

The term 'freshwater' itself can refer to a variety of environments. One of the most obvious differentiating factors among freshwater environments is the condition of water movement ("current"); i.e., static (so-called lakes) and running (rivers and streams). This factor can have primary affects on the biological community, e.g. organisms in rivers have morphological and ecological characters to stand against the running water, while many species exhibit adaptation to static and open water, especially in large lakes. It may also determine sampling methods: rapid currents will prevent setting nets, whereas, where there is no current, benthic organisms and bottom materials can be primarily collected with a hand net. Another factor is the transparency of the water, which significantly affects the light conditions, and hence the biocommunity structure, on the littoral bottom. This is also an important factor in the selection of sampling methods; scuba diving and other direct hand sampling are possible in clear waters, while indirect sampling methods are necessary in turbid waters.

Taking these factors into account, we apply a simple categorization of freshwater environments for the IBOY-DIWPA project: clear and turbid lakes, and slow and rapid rivers (Table 3.1). This classification is primarily based upon methodological considerations.

Practically, water depth is also a notable factor, as it limits the available sampling methodologies; hand collecting is possible only in the shallowest areas, and diving collection is recommended only in places less than 21 m deep. The bottom condition is another important factor in selecting the sampling method, e.g., bottom sampling is generally applicable only on soft bottoms.

Researchers of biodiversity must be aware of seasonal issues. For example, a marked fluctuation in water level is apparent in monsoon areas where a vast area of temporal freshwater appears during the flood season, while the water surface is covered by ice during the cold season in boreal areas. Therefore, we should select the best season for IBOY-DIWPA activities when monthly or frequent surveys are not practicable.

There is an intermediate habitat between fresh- and marine-waters where the salinity and water level conditions are greatly influenced by tidal activity. These environments can be researched using the methods outlined in this manual.

## Lakes

There are a variety of freshwater lakes harboring unique biota in the DIWPA region, i.e., the western Pacific and Asian region. Among them, Lakes Baikal and Biwa are well-known examples of "hotspots" of rich and unique biodiversity. Even for these lakes, though, despite considerable data accumulating about offshore plankton, there is little information on biodiversity in the littoral zone. Fortunately these are "clear" lakes, where direct sampling with the aid of scuba diving is practical. There are many other "clear" lakes in the DIWPA region; i.e., Lake Khuvsgol in Mongolia, Lake Toba in Sumatra, Lake Poso in Sulawesi Indonesia, and Lake Taupo in New Zealand. The IBOY-DIWPA project aims to use standardized methodology in order to compare the biodiversity in various clear lakes.

In addition to these "clear" lakes, there are many small and shallow lakes with turbid water, typically located in the lowlands (floodplain) formed by large rivers. Turbid lakes, including oxbow lakes, are regarded as one type of

*Table 3.1: Freshwater environments in East Pacific Asia and recommended methods for biodiversity investigations.*

| Biotic item | Lake | | River/stream | |
|---|---|---|---|---|
| | Clear lake | Turbid lake | Slow river | Rapid river |
| Fish | 1 electric shocker<br>2 net fishing<br>3 diving | | 1 electric shocker<br>2 net fishing | |
| Benthos<br>(Macro/Micro-Epiphyte)<br>(Macro/Meio/Micro-benthos) | 1 diving<br>2 hand collection<br>3 bottom sampler | | 1 hand collection<br>2 bottom sampler | |
| Plankton<br>(Zoo/Phyto-plankton) | 1 plankton net<br>2 bottle sampler | | 1 plankton net<br>2 bottle sampler | n/a |

freshwater habitat with a high variety of fish, especially in tropical regions. Despite the low transparency in these waters, the conventional methods for limnological/hydrobiological surveys listed in this manual can be applied.

**Rivers and streams**
There are more variable environments to be found in running waters (from small highland streams to large lowland rivers) than in static lakes. However, due to large fluctuations in their environmental conditions, e.g., resulting from seasonal and sporadic rainfalls, the observation of biodiversity in running waters is often complicated by the difficulty of choosing stable or ordinary conditions and selecting typical habitats comparable to elsewhere.

On the other hand, fauna and flora in lakes are more or less related to those of the inlet and outlet rivers by the migration of organisms. Thus, comparable biodiversity observations are also recommended for the respective in and out flowing rivers of lakes selected as observation sites for this project.

## 3.2.2 Target organisms
Amidst all of the various organisms in lakes and rivers, *fish, benthos,* and *plankton* are the primary targets for the IBOY-DIWPA project, selected on the basis of their ecological role, their dominance in the biological community, the pronounced variety of species diversity, and the availability of existing taxonomic knowledge and sampling methods.

**Fish**
Fish is selected as one of the main target taxa for the IBOY-DIWPA project because:
1) in many lakes, fish are ecologically dominant and influential organisms in the ecosystems,
2) qualitative sampling of fish is relatively easy via several collection methods,
3) quantitative or relative evaluation of abundance can be estimated through consideration of the ease/difficulty of catching them,
4) species composition in a given locality reflects environmental and zoogeographical factors,
5) fish are the most important protein resources from the lakes and rivers, and
6) ecological information (e.g., spawning season, growth patterns, etc.) can contribute to efforts to control the exploitation of freshwater resources by fishery and other activities.

We also attempt to collect information on fishery resources, such as fishes, shrimps and mollusks. At the initial stage of investigation, we recommend surveying the availability of information on target species from the fishery operators, whether cooperatives/companies or local fishermen. At the same time, analyze fishing methods/practices in an attempt to establish a proper measure for qualitative evaluation of fish diversity in the littoral zone of lakes.

## Benthic organisms

*a) Large benthic organisms*

The littoral zone is inhabited by a variety of organisms. Among them, large bottom-surface-dwelling (epi-benthic) organisms (macrobenthos including benthic fish and macrophytes) have been selected as the main target groups. These organisms are favorable for the evaluation of biodiversity due to the ease of collection, sorting, preservation and identification, and therefore, for the comparison of biodiversity between different habitats within a given freshwater body and between freshwater bodies with the same habitat type.

The macro-epibenthic organisms selected for the IBOY-DIWPA comprise mollusks (Gastropoda and Bivalvia), large-sized arthropods (Decapoda, Amphipoda, and Insecta), and submerged macrophytes (mainly vascular plants) for the following reasons:

1) these are ecologically dominant in most freshwater environments,
2) quantitative sampling is relatively easy due to their large size,
3) specialized treatment in the field is not necessary for the later identification of these groups,
4) existing alpha-level taxonomy of these groups is of a high standard, including readily available references for identification by non-specialists, and
5) they are attractive to the general public owing to their variable morphology and ecology. This characteristic enhances their capacity to raise public concern and is therefore important for continuing monitoring efforts into the future.

*b) Small benthic organisms*

The diversity of smaller benthic organisms, i.e., meio- and microbenthic species, have much greater diversity than the larger species. Thus, especially for inventorying purposes, the examination of these smaller organisms will be most informative and necessary. As they are more sensitive to environmental changes, they may serve as an indicator of environmental changes over short time-frames. Therefore, in parallel with the collection of macrobenthos, we recommend the inclusion of both qualitative and quantitative sampling of the smaller benthic organisms.

**Planktonic organisms**
There is no visible physical structure in the water column. However, planktonic microorganisms – comprising bacteria, phytoplankton, and zooplankton, etc. – are important components of biological communities with nutrient cycles, sustaining larger-sized animals at the higher trophic level. Among them, phytoplankton plays an important role as a primary producer in the aquatic ecosystem as well as plants on the terrestrial ecosystem. Although freshwater planktonic flora and fauna include many cosmopolitan species, environmental conditions in respective water bodies may largely affect their proportional abundance and species combination.

## 3.3 Field and laboratory methods and data management

### 3.3.1 Bottom types and site selection for the littoral bottom survey

Physical conditions in the littoral zone seem to be much more site-specific (variable from site to site) than those in the pelagic zone because of their high dependence on bottom characteristics. Site selection for the littoral survey therefore critically influences the results of observation. Priority in site selection should therefore be given to the littoral survey; the pelagic survey will then be conducted in the neighborhood of the littoral survey sites.

A variety of ecotones are recognizable in the littoral zone; e.g., from the steep rocky slope or the vertical rocky cliff to the gentle or flat slope with soft bottom often associated with macrophyte vegetation.

Our preliminary investigation shows that the rocky/pebble bottom (preferably, interspersed with sandy patches) is the most heterogeneous and most species-rich habitat in the littoral zone. Dense macrophyte vegetation is often developed to form a vegetation zone in the littoral zone, where wave actions are less influential owing to the site location or to water depth, and may therefore provide an important species-rich habitat. It is therefore recommended that rocky/pebble bottoms and macrophyte vegetation be selected as sites for littoral zone surveys. However, practical accessibility to the survey sites and availability of facilities are also important operational factors in determining research sites.

### 3.3.2 Water depth

Water depth is an important parameter for the selection of actual sampling sites in the littoral zone, due to its influence on light conditions and water temperature. This parameter also affects practical limits to the choice of sampling methods.

Underwater light conditions sufficient for photosynthesis are generally assumed to be at a depth 3 times the Secchi disk transparency; i.e., in the case of 5 m Secchi disk transparency (as in "clear" lakes), the compensation depth is expected to be 15 m. The maximum sampling depth should be planned primarily in consideration of the compensation depth or water transparency; i.e., in cases where the compensation depth is less than the depth range for safe diving, the depth range of the littoral survey should incorporate the compensation depth.

The surf zone is another unique habitat inhabited by organisms with life styles specialized for an aquatic-terrestrial transitional ecotone.

Many freshwater bodies (both lakes and rivers) are characterized by considerable fluctuations in water level, influenced by seasonal rainfall and water usage. We recommend sampling during the low water season.

Most macrophytes and benthos seem to be confined to the zone shallower than the thermocline depth (it is approximately 12–15 m deep in Lake Biwa), though some unique organisms may appear in the zone deeper than the thermocline.

Thus, it is recommended that sampling points be set at several depths, depending on the transparency and thermal conditions; e.g., 0–0.5 m (surf zone), 3–5 m, 7–10 m (above the thermocline), 15–20 m (below thermocline and around the compensation depth).

Since the water level of lakes is usually subject to seasonal fluctuation, these depths should be measured in comparison to the "standard water level" for the respective water bodies.

### 3.3.3 Transect line

In order to observe the biodiversity patterns at each site for the littoral survey, it is necessary to obtain a set of data and samples in various depths and substratum types. For this purpose, we recommend setting a transect line perpendicular to the shoreline. The transect line is helpful as a guideline to figure out the rough distribution patterns of substratum types and of macrophyte vegetation (or coverage of sponges) along a depth gradient while describing the bathymetric profile of the survey site.

For surveys from the water surface, the transect line should be marked using anchored buoys. If scuba diving is to be used, the transect line should be marked using a brightly coloured (white or yellow) measuring tape or a rolled ruler with distance marks at 5 m intervals. The starting point of the transect line (i.e., the shore-side end of the line) should be precisely located by easily recognizable landmark(s), and preferably with GPS data (longitude and latitude to the nearest second), in order to correctly relocate the transect for future surveys.

In addition, the distribution patterns of substratum types (and macrophyte vegetation and/or coverage of sponges) along the bathymetric gradient should be recorded.

### 3.3.4 Physical structure and underwater landscape

Substratum type and water depth are significant environmental factors for micro-distribution of each organism. These should be carefully recorded to describe the habitat.

The depth profile along the transect line should be described by measuring the water depth with an echo-sounder or by direct measurement at regular intervals along the line or by recording the distance along the line for a given water depth. An echo-sounder can provide an accurate profile of the bottom for depths greater than approximately 0.8 m as well as images of the development of macrophyte vegetation.

Substratum types can be largely categorized into:
1) bedrock or rock (having surface area much larger than the quadrat area employed),
2) stone (less than approx. 20 cm on the longer axis),
3) gravel (stony particles smaller than approx. 5 cm on the longer axis),
4) sand, and
5) mud or silt.

Considering the actual collection of benthos (see below), a fixed area (quadrat) sampling is available on substratum categories 1), 3), 4), and 5), whereas "stone-unit sampling" is preferable for category 2). Since the categorization into 3), 4) and 5) is subjective, we recommend collecting additional samples for particle size analyses.

In the case of clean lakes and rivers, underwater photographs (videos) are also useful in addition to the biological samples.

### 3.3.5 Water column survey

The littoral survey of the present IBOY Project should involve surveying biota in the water column extending above the substratum in the littoral zone. The environmental variables and the community of small planktonic organisms in the water column are typically structured vertically. Therefore, researchers should collect water samples from varying water depths depending on the transparency and thermal conditions of the water column, using site selection as discussed above for the bottom survey. We tentatively propose collecting samples from the water column at points along, or in the extended direction of, the transect line with bottom depths of 5 m, 10 m and 20 m in waters with sufficient depth. A portable GPS is useful for obtaining accurate latitude and longitude data in the field. If a GPS device is not available, they can be read from a topographic map. The most suitable topographic maps are the TPC series (scale: 1:500 000) which covers worldwide land topography.

Biological samples for bacteria, pico-, nano-, and micro-phytoplankton, heterotrophic nanoflagellates, and ciliates require preservation via special fixation procedures.

### 3.3.6 Physico-chemical variables

Temperature and other physico-chemical variables in the water column are important environmental conditions. These should be measured along the depth gradient when and where sampling is conducted. However, since these variables are subject to temporal fluctuation, the compilation of long-term records on these variables using data-logging devices is desirable.

The chemical variables in the environment may include dissolved oxygen (DO), pH, nutrients (compounds of nitrogen and of phosphorus), silicate, dissolved organic carbon (DOC), and particulate carbon/nitrogen/phosphorus. These variables are measured from sub-samples of the water sample. Vertical profiles of water temperature and electrical conductivity are measured with a portable multi-profiler. Light condition is estimated from the Secchi disk reading, etc. An underwater quantum meter is the most effective means for estimating underwater light conditions.

### 3.3.7 Data from meteorological stations and remote sensing

Data from a meteorological station located within the watershed of the survey site should be utlised to determine seasonal patterns of temperature and precipitation, which are highly influential in freshwater environments. In some lakes such as Lake Biwa, there are automatic observatories recording several conditional variables of the lake water (water temperature, DO, BOD, COD, transparency, etc). Such data are also useful for describing general patterns of the environmental condition.

For large water bodies such as Lakes Baikal and Biwa, regular low-cost NOAA /AVHRR data can be used as well as free SeaWiFS and EOS /MODIS data. These data have aerial resolution of 1.1 km in nadir, 5 spectral bands from visual to infrared for AVHRR, 8 visual spectral bands for SeaWiFS (the same spatial resolution). MODIS data has two bands with 200 m aerial resolution, which is quite good for lake studies, three bands with 500 m resolution, and 17 bands with 1 km resolution. EOS/ASTER data have 90 m resolution for infrared band and 15 m resolution for visual-near infrared band and is quite good for limnological studies.

Such limnological parameters as water turbidity, surface temperature, reflectance in chlorophyll, dissolved organic matters and sediment optical bands can be derived from imagery using standard algorithms, included in some free (SeaDAS) and various commercial software packages. SeaWiFS and MODIS data is available free from NASA, USA for registered users, however, SeaWiFS images can be browsed on-line, which is useful for choosing good quality images. MODIS provides a huge volume of data available on 8 mm video tapes. High resolution ASTER data is available from Japan Earth Remote Sensing Data Analysis Centre (ERSDAC) by individual application.

### 3.3.8 Basic unit for field investigation

An important purpose of the present IBOY Project is to compare the biodiversity between freshwater bodies and to record the present situations for initiating long-term monitoring at respective freshwater sites. For surveys of freshwater as proposed here, it is ideal to have well-equipped research vessels, and experienced scuba divers (Fig. 3.1). In order to obtain comparable field data between different freshwater sites, we must consider the minimum unit for investigation, and anticipate the least equipped situation.

**Research vessel**

A research vessel is fundamentally necessary for surveying freshwaters. In the case of water column sampling, most sample and data collection is done onboard the vessel (and partly from the shoreside). For bottom surveys conducted by scuba divers, the divers must cover a long distance where the lake bottom slope is gentle enough. A surface vessel can provide assistance to the divers while tracing their movements underwater. It is ideal to have a specially equipped vessel for limnological research. However, a wooden or rubber boat with an outboard engine can be an adequate research vessel for the present purpose.

Onboard the research vessel, one person should be fully engaged in operating the vessel (the "captain"). Ideally, two people should be engaged as onboard research staff to obtain and record data: one to operate sampling gear and read data and the other to record the data read by his/her partner ("onboard staff"). These two persons can also carry out the collection and treatment of samples: one operating the samplers and the other treating collected samples. Therefore, the minimum number of people required to staff the research vessel is three, including one who is licensed to operate the vessel.

Scuba diving activities require at least two experienced scuba divers ("underwater staff"). In this case, the on-board staff will be free from most of the sample collection, and can focus on the treatment of samples collected by the underwater staff. Thus, the minimum unit for field surveys employing scuba diving is composed of five persons (one captain, two on-board staff, and two underwater staff).

In either case, additional members on board (dependent upon the capacity of the vessel) enhances the unit's capacity to collect and treat data and samples.

**Scuba divers**

Where the environmental conditions are suitable for data collection by scuba diving, surveys require scuba divers who are skilled in underwater sample collection. It is much more important that they be experienced researchers than professional divers. Scuba diving also provides a good opportunity for researchers to make direct observations of the underwater landscape and biota, thereby extending and deepening their understanding.

For safe scuba diving operations, buddy diving (a pair of divers cooperatively engaging in underwater activities) is absolutely essential. If resources permit, some onboard staff can accompany the underwater staff to assist with sample collection.

### 3.3.9 Data management

The data on each specimen will be maintained in an electronic database (available on the internet for researchers). The following information for each unit-sample is required for constructing the database:
1) catalog (sample) number,
2) species name(s),
3) genus name(s),
4) family name, and family cord (see appendix),
5) order name,
6) number of individuals in the sample,
7) size of smallest and largest individuals (or individual size list),
8) weight of smallest and largest individuals (or individual weight list),
9) locality,
10) latitude and longitude,

# Freshwater Ecosystems

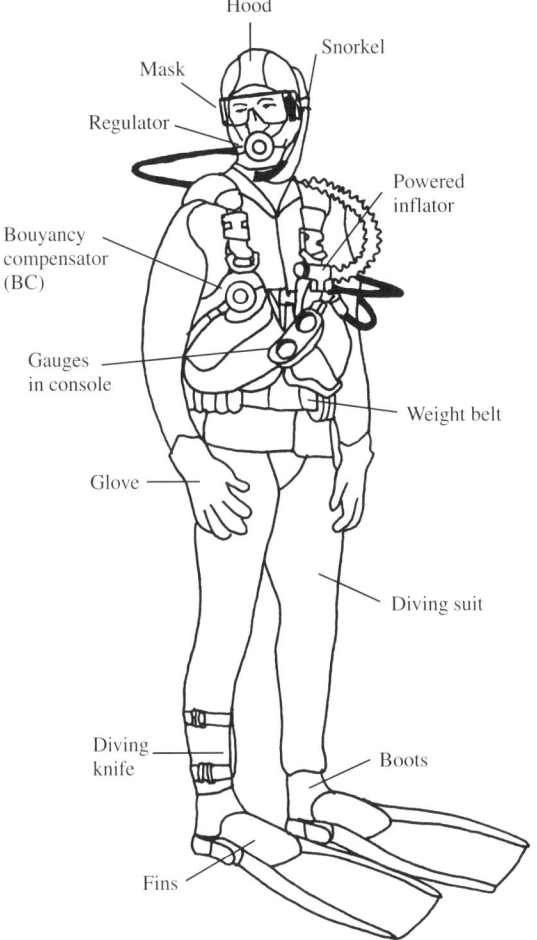

*Figure 3.1: Scuba diver with diving gear (after PADI Open Water Diver Manual, 1999)*

11) habitat,
12) water temperature (at the surface, 1 m depth, and bottom)
13) water pH (at the surface, 1 m depth, and bottom)
14) water DO (at the surface, 1 m depth, and bottom)
15) water depth at collecting point (at the beginning and end of collection),
15) time of collection (at the beginning and end of collection),
16) date of collection,
17) collecting method,

18) name of collector(s),
19) fixation and preservation methods
20) name of identifier(s), and
21) remarks (some comments on specimen, if present).

See 3.4.1 for field sampling methods and Table 3.2 for a sample Field Data Sheet.

## 3.4 Biodiversity survey for each organism

### 3.4.1 Fish

In addition to the biodiversity data in fishes, we recommend accumulating data on 1) diurnal behavioral rhythm, 2) seasonal migration patterns, 3) diet of each species, 4) spawning season, and 5) population size.

**Field Sampling**

Fish are the most mobile animals in the freshwater ecosystem. Although some fish have strong site-tenacity, remaining on the same site for a long time, many other fish exhibit a high mobility relating to the seasonal and diurnal migrations. Collection of fish samples should take account of their high mobility. Stationary collection gear such as gill nets and electrofishing are suitable for this application.

"Fish occurring" will be clarified by the survey based on diurnal behavioral rhythm, seasonal distribution and migration patterns of the collected fish.

Studying "diurnal behavioral rhythm" requires 24 hours of sampling: fish are collected from each gill net every 3 hours, e.g., 12:00, 15:00, 18:00, 21:00, 24:00, 03:00, 06:00, 09:00, 12:00, in the ideal case. If the interval between collections is shorter than one hour, the collected fish samples should generally provide good data for food habit analysis.

"Seasonal distribution and migration patterns" can be investigated by periodical surveys using gill nets. Samples should be collected three times in two weeks, with four two-week collecting periods in a year, viz., the beginning of the rainy season, middle of the rainy season, end of the rainy season, and middle of the dry season in tropical areas; four times in temperate areas, viz., spring, summer, autumn and winter; and three times in boreal areas, viz., spring, summer and autumn.

There is a variety of equipment available for collecting fish, targeting fish with diversified habitat preferences and behavioral properties. For surveying inventories of fish with high diversity, several collecting methods may be employed to supplement each other. Here, we review several representative methods for the IBOY project.

*a) Gill net* (Fig. 3.2)

A gill net is a portable, stationary fishing device and one of the most effective tools for inventorying fish fauna. A gill net consists of a long rectangular (belt-like) net sewn with semi-transparent strings, with floats at the top and weights at the bottom of the net. It is set on the bottom and stands vertically owing to tension between the floats and weights. Gill nets capture fish when the fish head hits the net and penetrates the mesh but cannot pass through the mesh. Obviously, gill nets are very specific with regards to the body size (body height) of fish.

For the IBOY project, a standard set of gill nets should include five different mesh sizes, each 2 m high and 40 m long with mesh sizes of 3.0 cm, 2.5 cm, 2.0 cm, 1.5 cm, and 1.0 cm. These types of gill nets are expected to catch respective sizes of fish. Each net has a steel chain affixed to its bottom as a sinker. By varying the weight of the chain, the depth of the gill net can be set, viz., surface or bottom.

When collecting fish with scuba, the divers set a screen net, tracing three sides of an imaginary quadrat (1 m × 1 m, 2 m × 1 m, etc., dependent on the densities of target fish). Special attention should be paid to ensuring the net fits close to the bottom; there should not be any space to allow fish to escape under the net. In stony bottom areas, the weight code or chain of the net should be placed under the stones that are located on the border of the quadrat. Then, the

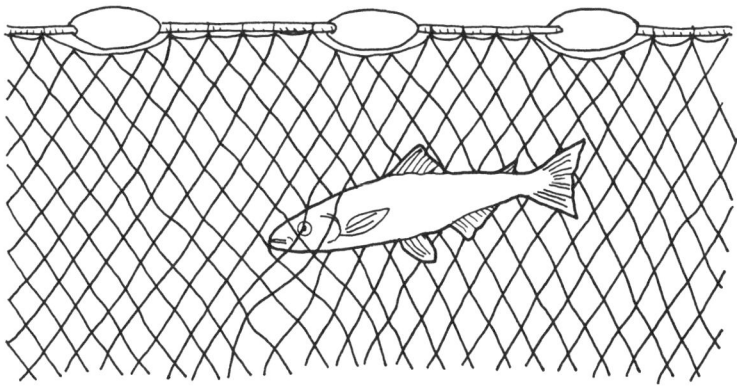

*Figure 3.2: Gill net (after Inoue, 1983).*

fish within the quadrat area are chased to the screen net and captured in a hand net. Where there are loose stones within the quadrat area, each stone should be turned in order to flush out any fish hiding behind the stones. The use of chemicals such as rotenone may improve the efficiency of collecting fish where permitted.

*b) Electrofishing*

Electrofishing is effective in researching fish fauna and populations in shallow areas (e.g., a mountain stream). Three people are required to conduct electrofishing; one to operate the electro-shocking gear and two for collecting fish with hand nets. The effective area for electrofishing is 50–100 m. To avoid sampling errors during electrofishing, 5 mm mesh seines should be used to surround the sampling site. We recommend the Smith-Root Model 15 electro-shocking unit (300-500 V, AC 90 HZ; Vancouver, Washington, USA).

For analyzing the fish population, three-pass depletion-removal electrofishing is recommended. The following method for three-pass electrofishing in streams is cited from Li & Li (1996) with some modifications.

1) Block the upstream and downstream margins of habitat to be sampled with block seines. Be sure to weight the bottom of the net with rocks to prevent fish from escaping.
2) Using an electro-shocking gear make three passes along the length of the blocked stream segment. More than 30 minutes is needed between passes.
3) After each pass, put fish into a bucket or basket marked with a pass number and the number of each species collected. This approach assumes that less fish will be caught at each successive pass given equal collecting effort. If the number of fish does not decrease, six passes will need to be conducted, viz., first + second, third + forth and fifth + sixth passes are treated as first, second and third passes, respectively.

For the methods of population estimate, see Riley & Faush (1992).

*c) Other collecting gear*

*Minnow trap* (or cage trap. Fig. 3.3): A kind of cylindrical cage, formed by weaving slender pieces of wood or bamboo stems in parallel. At one end of the trap, the slender pieces are tied to make a "dead end". At the other end, a conical part made of similar pieces is fitted with its apex directed to the inner part of the cylinder. The apex of this cone has an opening, through which fish can enter the trap.

Before setting a minnow trap, some bait should be prepared such as pieces of fish meat or wet paste of powdered bait (made of dried cocoons or potatoes,

# Freshwater Ecosystems

*Figure 3.3: Minnow trap. (after Yuma et al. 1997)*

etc.) and placed in the trap to attract fish. Experience suggests that this trap should be placed directly on the substratum.

Minnow traps have the advantage of collecting fish without injury, but there are also at least two disadvantages: 1) it can be used only in bodies of water with very little or no water current, and 2) the species collected are significantly influenced by the type of bait used.

*Casting net*: A cone shaped net whose diameter is about 4 m, with a rope connected to the top and an anchor chain attached to the base. The mesh size of the net is 1.0–1.5 cm.

While holding the rope the net is cast on the water's surface every few meters. When properly cast, the base of the net expands in a circle to surround an area some meters across and sinks to the bottom. Then, as the rope is pulled, the base of the net gradually closes while moving across the bottom, and finally, fish within the area are captured in the net.

A casting net is one of the most useful devices for collecting fish in shallow water, both in lakes and rivers. When wading in a stream to use this net, we must work from the lower to the upper reaches to minimize the effect of noise on the fish.

With this net, most of the larger fish within the area of the net can be collected. Since the coverage area of the opening of the net is given, fish abundance can be easily calculated. However, this is only practical in open areas with smooth bottoms.

At least ten casts are required to collect enough fish at each site.

*Seine net*: An encircling large fish net, held upright in the water by floats along the top and weights at the bottom. It is hauled to the shore or to a boat, and useful for collecting fish including smaller individuals inhabiting sandy or muddy bottoms without hard obstacles such as stones and tree twigs.

A seine net operation requires at least two persons to pull the ropes on each side of the net. This net is usually drawn (seined) from about one meter depth in an area adjacent to a bank. A 10 m long × 2 m high net with a mesh size of 2.0–10 mm is convenient to operate and to collect fish.

*Hand net* (Fig. 3.4): A round-shaped or D-shaped hand net with a long handle is useful for collecting small fish from near the surface or interstitial spaces of stone or sand. The frame diameter is generally 30–50 cm with a mesh size of 2–5 mm. The depth of the net is 1.5–2.0 times the diameter of the frame. Mesh sizes less than 5 mm are useful for collecting juveniles or small species, but the small mesh size makes quickly scooping up fish difficult. Small fish and juveniles in floating plants can be scooped with the plants by net from the under side. To collect fish from interstitial spaces of stone, sand or bush, a D-shaped net is effective. The frame size is dependent on the sampling point. The straight side of the frame is located on the bottom without gaps. Using a hand or foot, fish are chased into the net from the interstitial space of stone, sand or bush. Hand net sampling must be done after casting net sampling.

In clear and shallow areas, snorkel diving with a small hand net is very effective for collecting small and bottom dwelling fishes.

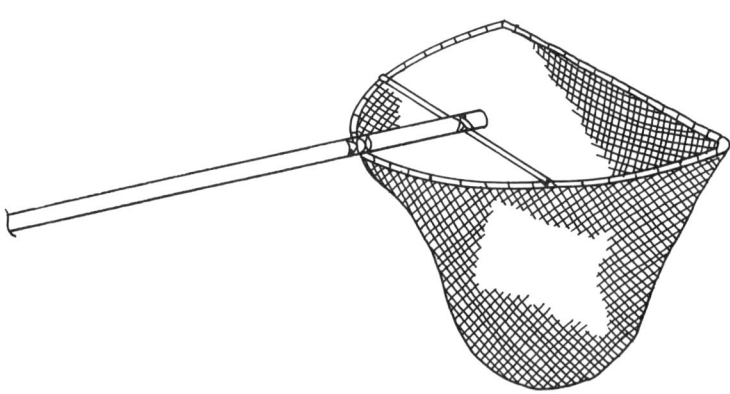

*Figure 3.4: D-shaped hand net. (after Handa et al. 1987)*

*Screen net*: It is generally difficult to collect smaller-sized demersal fish with less mobile habits by the methods mentioned above. Scuba divers using a screen net and a hand net can collect fish such as gobies and sculpins.

Screen nets have a similar shape to gill nets, with floats at the top and weights at the bottom. The length of the net is determined by the size of quadrat area to be surrounded by this net. The mesh size of the net (screen) is much finer than for gill nets, i.e., a few millimeters across, in order to collect smaller fish. The screen is almost vertically set underwater owing to the tension generated by the floats and weights. However, a higher tension is necessary for the screen net than for gill net because the screen part receives much larger force from the water current, which often makes the screen lay almost flat on the bottom. Thus the height of the net should be determined by the force from water current but be sufficient to prevent fish escaping midwater, at least higher than 50 cm. The weight of the net is usually a steel chain or a code (cloth wrapped around a lead core), which is flexible enough to maintain close contact with the bottom surface.

## Field treatment
### a) Sampling lots and labelling
Field assortment of sampling lots is preferable for both fish samples and samples of other organisms. For example, fish samples collected with multiple types of gill net should be stored and labelled as different sampling lots in the field, with the aim of producing important information about the size selectivity of each type of gill net.

Specimens of fish from different sampling points should be carefully placed in separate containers. A label to distinguish sampling points and methods should be placed in each container. The following information is necessary on each label (see the "Field Data Sheet", Table 3.2):

1) Catalog (sample) number.
2) Name of lake/river.
3) Latitude and longitude measured by a portable GPS system or a topographic map.
4) Location of sampling point (including name of the nearest village, and its distance and direction).
5) Methods of collection (types of gear, amount of operational effort, etc.)
6) Water depth
7) Substratum type
8) Starting and ending time/date of collection
9) Name(s) of collector(s)

10) Weather
11) Air temperature
12) Water temperature
13) Species included
14) Approximate quantity of collected organisms
15) Remarks

This set of data should be recorded on water-resistant paper with a waterproof pigment ink or soft pencil. Avoid using gel, or other types of pens that easily and quickly fade from exposure to sunlight, etc.

*Table 3.2: Sample field data sheet.*

| **FIELD DATA SHEET (COLLECTION LOT)** | | **IBOY/Freshwater 2001** |
|---|---|---|
| **Name of lake/river** Lake Biwa | **Catalog (sample) no.** | IBOY/FW-Biwa-010901-01 |
| **Location of collection site** | | |
| Longitude/latitude (measured by GPS) | 135°12'34.56" E / 35°22'33.44" N | |
| Location (name of the nearest village, distance and direction from it, etc.) | Southern coast of Okude-Bay, between Okura and Sugaura villages, northern basin of Lake Biwa, Shiga Prefecture, Japan | |
| Remarks | | |
| **Description of collection** | | |
| Methods (type of gear, amount of operational effort, etc.) | Bottom sample, 25 × 25 cm quadrat sampling by scuba, 0.5 mm mesh hand net | |
| Water depth | 5.0 m | |
| Substratum | Sandy-pebble with macrophytes | |
| Starting-ending time /date of collection | 14:30-15:20, 2001 Sep 01 | |
| Name(s) of collector(s) | K. Nakai, M. Yuma | |
| Remarks | | |
| **Field conditions** | | |
| Weather | Fine and windy | |
| Air temperature | 28°C | |
| Water temperature | 28°C | |
| Remarks | | |
| **Rough contents of sample** | | |
| Species included | Biwamelania spp., Unio biwae | |
| Approx. quantity | ca. 50 individuals | |

*b) Fixation for preservation*

Samples of collected fish should receive proper fixation for later examination in the laboratory and for long-term preservation. Generally, fish are relatively large compared to other freshwater organisms, and have large abdomens full of viscera. Thus, a strong fixative is required for fish.

Specimens must be fixed by 10 % formalin solution (40 % aqueous solution of formaldehyde:water = 1:9). To fix a specimen in good condition, a quantity of formalin solution greater than five times the weight of the specimen is required. Small fish (fresh or alive) should be placed directly into a bottle filled with formalin solution immediately. If the fish are alive, their fins are usually extended sufficiently and properly fixed by this treatment. Large fish (more than 15 cm in total length) should have pure formalin injected into the abdominal cavity to ensure good fixation of the viscera, and then be placed into a container of formalin solution. In specimens more than 30 cm in length, the right side of the abdomen should be cut and the dorsal muscle should be injected with pure formalin to allow a prompt fixation.

*c) Photography*

Coloration is one of the important keys for identifying fish species. However, fish coloration is usually lost after preservation in any fixative. Although many species can retain their live coloration for a while even in a formalin solution, coloration in some species change immediately after death,. It is therefore recommended that pictorial data of live fish be recorded by photographs or video, since coloration (color and patterns) is very difficult to properly describe in words.

Photographing fresh specimens in the field requires: a single-lens reflex camera, a macro lens, a release, tripod, a focusing stage and a white board as a screen. Color slide film (ISO 100) is preferable to print film. A digital camera with high resolution can also be useful for prompt recording. It is advisable to include a standard color card in each photograph (Fig. 3.5) to justify the colors.

Photographs should be taken in shade to avoid any shadow from the specimen and reflected sunshine on the body of the specimen. If there is no shade naturally available, an umbrella can be used for this purpose. Use the following protocols for taking photographs:

1) Set-up a single-lens reflex camera with a macro lens, release, tripod, focusing stage and a white board.
2) Put the specimen on the white board, left-side up with a photograph number and a scale.
3) Set the f-stop on the lens to f-11 or f-16.

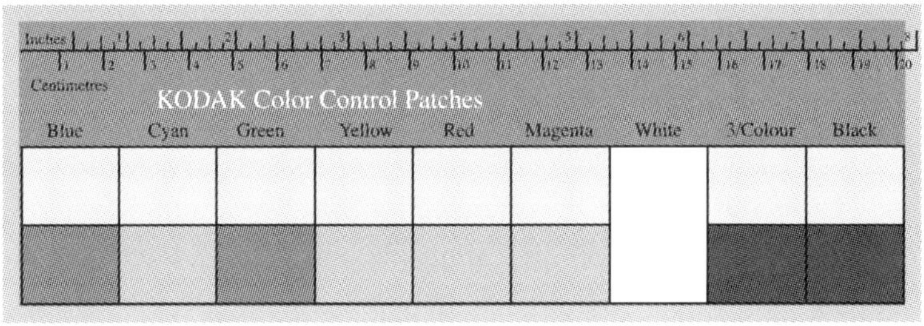

*Figure 3.5: Color control patches.*

4) Take four photographs of each specimen; one at normal exposure, then +1/3 over, +1 1/3, and +2 by using the adjustment dial for exposure (these adjustments are for Nikon. If your camera differs, please see the manual).
5) Photographed specimens should be separated from other specimens and labelled with collecting data and a photograph number.

## Laboratory procedures

*a) Primary fixation*

Fish sample lots that are separately stocked with formalin after collection in the field should be put into a new 10 % formalin solution within two weeks after collection for complete fixation. These specimens also should be kept in a cool and dark place.

*b) Long-term preservation and permanent labelling*

For long-term preservation, sample lots should be divided into species lots after identification of species (or species group). Each new lot (preferably by species) should be labelled with a new catalog (sample) number that refers to the IBOY/ Freshwater Field Data Sheet (see 3.3.9).

During this treatment the fixative solution should be replaced. Since some disadvantages of using formalin have recently been recognized, it is recommended that fixed samples be preserved with another medium such as ethanol solution. But ethanol is very expensive and highly flammable, and there may be some legal regulation concerning the storage and handling of large quantities of ethanol. Conditions permitting, we recommend replacing the preservation medium with ethanol. However, as lipid contents may be gradually lost during preservation in ethanol, taxonomic and ecological measurements should be conducted during this procedure (see 3.4.1).

When fixation has been achieved with formalin, specimens of each sample lot should be soaked in a tub of water (with a label attached to each field lot) in order to remove the formalin. The water in the tub should be continuously replenished to affect the removal of formalin. It will take about 24-hours to remove formalin sufficient for later long-term preservation and storage. After the formalin has been removed, specimens should be placed in a 70 % ethanol solution for long-term preservation.

If there is difficulty in obtaining or using large quantities of ethanol, the specimen can continue to be preserved in formalin until conditions improve. Only neutralized formalin (or formalin solution with marble tips) can be used for long-term preservation.

## Identification

There is no comprehensive book for identifying Indonesian freshwater fishes. Although Kottelat et al (1993) provides useful pictures for first-stage fish identification, we must compare recent descriptions of species for critical identification. Recent papers on Southeast Asian fish are primarily published in The Raffles Bulletin of Zoology, Ichthyological Exploration of Freshwaters and The Natural History Bulletin of the Siam Society. Publications by Weber & de Beaufort (1911, 1913, 1916, 1922, 1929, 1931, 1936, 1940, 1951, 1953, 1962) are also useful. Taxonomic studies on cypriniform fishes of Southeast Asia were reviewed by Doi (1997).

Masuda et al. (1984) and Nakabo (1993) are useful guides for identifying Japanese fishes. The former contains fine photographs of fish while the latter contains an order to species identification key form.

Taliev (1955), Berg (1965), Smirnov & Shumilov (1968), and Sideleva (1982) are useful for the identification of Russian fish, especially of Lake Baikal.

## Counts, measurement and ecological data

Count and measurement methods for fish specimens usually follow Hubbs & Lagler (1974). Here, we describe count and measurement methods that are somewhat rearranged according to recent ichthyology.

*a) Counts*

*Number of dorsal fin rays (D)*: Dorsal fins are classified according to their number: one dorsal fin type (e.g., Cypriniformes) and two dorsal fins type (e.g., Perciformes). When there is one dorsal fin, the dorsal fin contains unbranched and branched rays. Ray counts are given in the following sequence: "number of unbranched ray/number of branched ray" (e.g., 4/8). In some cyprinid and

Siluriform fishes, the last unbranched ray is strongly osseoused. For the count of unbranched rays, because small-unbranched rays can not be seen from outside, soft X-rays are required.

In two dorsal fins type the anterior fin consists only of spines and the posterior fin consists of one anterior spine followed by soft (branched) rays. Spines are designated by large Roman numerals (I, II, III...). Soft (branched) rays are designated by Arabic numerals (1, 2, 3...). Thus, a dorsal fin that consists of "ten spines and 14 soft rays" is designated as "X, 14".

*Number of anal fin rays (A)*: Anal fins are divided into two types; one consisting of unbranched and branched rays (e.g., Cypriniformes) and the other consisting of spines and branched rays (e.g., Perciformes). The former is designated by the "number of unbranched rays/number of branched rays" (e.g., 3/5) and the latter as "number of spines, number of branched rays" (e.g., III, 8). In some cyprinid, the last unbranched ray is strongly osseoused.

*Number of pectoral fin rays (P1)*: Although the number of pectoral fin rays is designated by the total number (e.g., 14), in species of Balitoridae the number of unbranched rays is very important for identification. In these species, the number of pectoral fin rays should be designated by the "number of unbranched rays/number of branched rays" (e.g., 8/16). In Siluriform fishes, unbranched rays are strongly osseoused.

*Number of pelvic fin rays (P2)*: Pelvic fins are divided into two types: one consisting of unbranched and branched rays (e.g., Cypriniformes) and the other consisting of spines and branched rays (e.g., Perciformes). The former is designated by the "number of unbranched rays/number of branched rays" (e.g., 1/7). The latter is designated by the "number of spines, number of branched rays" (e.g., I, 7).

*Number of caudal fin rays (C)*: Number of caudal fin rays is determined by the "number of branched rays in upper lobe + number of branched rays in lower lobe = total number of branched rays".

*Total number of lateral line scales (TLLS)*: The total number of lateral line scales is determined by the "number of scales along the lateral line scales" and "scales on caudal fin". The count of the number of scales along the lateral line scales terminates at the structural caudal base as determined by moving the caudal fin from side to side. A series of scales on the caudal fin extend from the end of the lateral line scales on the caudal fin lobe. "Number of scales along the lateral line scales" + "scales on caudal fin" = "total number of lateral line scales" (e.g., 32 + 3 = 35).

*Number of predorsal scales (PrDS)*: This is the total number of scales along a straight midline from the occiput to the origin of the dorsal fin.

*Number of scales between the lateral line and the origin of the dorsal fin, anal fin and pelvic fin (D-LLS, LL-AS, LL-P2S)*: In these scale counts, scales immediately in front of insertion of dorsal, anal and pelvic fins are counted as 0.5. Scales on the lateral line are not included.

*Number of circumpeduncular scales (CpS)*: This number is counted at the lowest part of the caudal peduncle.

*Number of gill rakers (GR)*: This number is counted from the first gill arch. The number is designated by the "number of the gill rakers on the upper rib/ number on the lower rib" (e.g., 8/30). Some dissection is required to count gill rakers.

*Number of vertebrae (Ver)*: Vertebrae are divided into abdominal vertebrae without an elongated haemal spine and caudal vertebrae with an elongated haemal spine. A urostyle at the end of the vertebrae is included in the caudal vertebrae count. The number of vertebrae is presented as the "number of abdominal vertebrae" + "number of caudal vertebrae" = "total vertebrae" (e.g., $16 + 17 = 33$). To count vertebrae without dissection, a soft X-ray photograph is required.

*Number of predorsal (PrDVer) and preanal vertebrae (PrAVer)*: Predorsal and preanal vertebrae refer, respectively, to vertebrae with neural (haemal) spines in front of the most anterior dorsal (anal) pterygiophore.

*b) Measurement*

    Standard length (SL)
    Head length (HL)
    Head depth (HD)
    Head width (HW)
    Body depth (BD)
    Body width (BW)
    Caudal peduncle length (CpL)
    Caudal peduncle depth (CpD)
    Predorsal length (PrDL)
    Preanal length (PrAL)
    Prepelvic length (PrP2)
    Length of upper caudal fin lobe (UCL)
    Length of lower caudal fin lobe (LCL)
    Snout length (SnL)
    Orbital diameter (OD)
    Postorbital length (PostOrb)
    Interorbital length (IntOrb)

*c) Ecological Data*
Using fixed specimens of fish, a variety of ecological data can be obtained; for example, body size (length and weight), food habit (from stomach/gut contents), sexual condition (gonadosomatic index = <gonad weight>/<somatic weight> × 1000), etc. For these analyses, we recommend following the procedures described above for transferring the samples from formalin to ethanol (see p.132) to remove excessive formalin from the fixed samples (for health and safety reasons).

Lipid contents in particular, may be gradually and considerably lost through preservation in ethanol, so the body weight should be measured before preservation in ethanol. When the stomach/gut contents or the conditions of gonads are examined, it is better to separate the digestive canals (stomach and gut) and gonads from the body by dissection and preserve them in separate vials with formalin solution. Of course, digestive canals and gonads should be labelled with details to specify the individual of their origin.

## 3.4.2 Macrobenthos and macrophytes
**Field methods**
A variety of methods for collecting macrobenthos in the littoral zone have been proposed (e.g., Kajak 1971). It is now recognized that the substratum types and topography in the littoral zone are extremely heterogeneous, and that conventional on-board samplers for macrobenthos collection are restricted to soft bottom habitats. Thus, for the IBOY Project, where scuba diving is available and the environmental conditions make it practical, we recommend employing it for direct observation and sampling of the littoral zone.

There are numerous types of underwater habitats in the littoral zone, from which we have selected several types according to the substratum conditions for biodiversity observation where qualitative sampling is possible using standardized procedures. We describe the methods of sampling for four kinds of typical underwater habitats prevailing in the littoral zone as follows:

a) *Unconsolidated bottom (mud, sand, fine gravel) without vegetation*: Establish a quadrat; scoop the bottom materials in the area by a hand net.
b) *Bedrock area*: Establish a quadrat; scrape and sweep the attached organisms off of the bedrock surface in the area.
c) *Macrophyte vegetation (usually developing on an unconsolidated bottom)*: Establish a quadrat; root-out macrophytes growing in the area by hand, and scoop the bottom materials in the area by a hand net.
d) *Loose (not submerged) stones*: Select a stone, collect it by hand, and sweep the exposed bottom after removal of the stone.

Beforehand, we describe quadrat area size and sampler mesh size, which have been devised in the course of actual experiences in the field, chiefly due to operational easiness of underwater manipulation. Rather than the standard hand net, it is possible to use a portable suction sampler operated by air bubbles supplied from the scuba system to collect samples. It is not practical, however, to use this device on silty bottoms where its operation may greatly increase the turbidity of the water.

## Quadrat size and sampler mesh
### a) Area to be sampled
Drawing on field experiences, we provisionally recommend that the area for sampling macrophyte vegetation and the unconsolidated bottom be around 25 cm × 25 cm (625 cm$^2$), giving due consideration to the following conditions:
1) Operational ease (i.e., smooth operation of hand net, avoiding collecting excessive bottom materials and macrophytes, etc.) and scuba diving time limitations, which limit the maximum practical area.
2) Low density of the targeted macrobenthos (e.g., mollusks and large-sized arthropods), which favours a sample area as large as possible.
3) The distribution and size of patches of vegetation and soft bottom in a rocky bottom area, which determines the maximum size of a sample area with uniform habitat.

The same sized area will be applied to sampling the bedrock surface, where macrobenthos attached to the rock surfaces can be swept off.

This area is the minimum size that will include a sufficient number of larger target organisms, especially mollusks and arthropods whose density is relatively low. Of course, researchers must have the flexibility to enlarge the area (50 cm × 50 cm, for instance) where the low density of target organisms results in an insufficient number of samples to be collected in a smaller area.

It is noteworthy that the proposed sample area presently is around twice the size of common bottom samplers (e.g., Ekman-Birge sampler [Fig. 3.6]) that can cover an area of ca. 15 cm × 15 cm or 20 cm × 20 cm. At the same time, the proposed area size is almost the practical maximum to: 1) easily scoop the whole bottom material with a hand net of operative size and 2) facilitate finding a suitable soft-bottom patch for sampling a rocky bottom where no large soft bottom patch can be found.

It should be noted that, on the one hand, the density of extra-large bivalves (e.g., some unionid mussels in Lake Biwa) is very low but significantly affects the biomass of a sample if included. We therefore recommend selecting sampling sites that do not include these "giant animals" whose density and biomass should be investigated by different methods covering a much larger area.

**Biodiversity Research Methods**

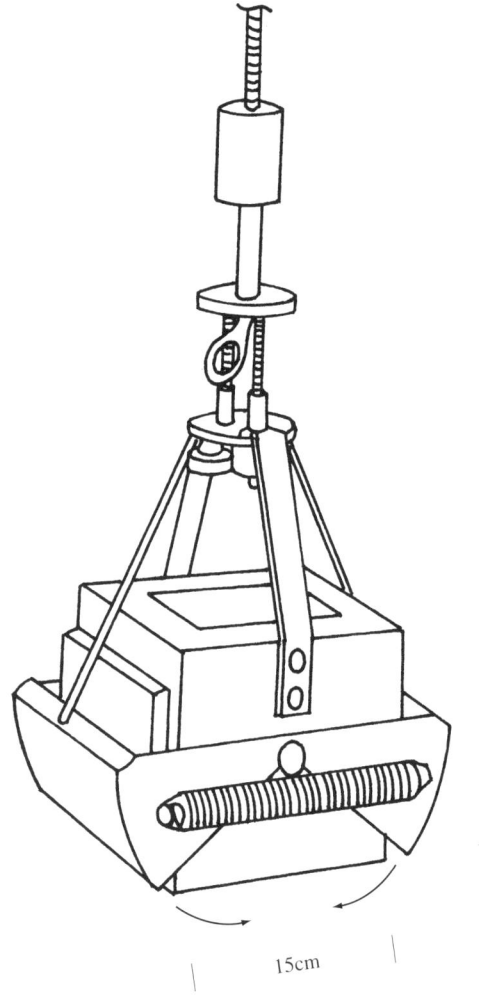

*Figure 3.7: Ekman-Birge sampler. (after Handa et al. 1987)*

The recommended area may, on the other hand, be too large for some small but dominant taxa such as oligochaetes and chironomids. Thus, these taxa should be collected from a smaller area. Standardized sampling area sizes for respective taxa will be determined during the preliminary phase of this project.

On a bottom where the size of substratum particles is excessively large, it is very difficult to sample the bottom of a fixed area because of the irregular shape of particles. Therefore, we provisionally propose to select a medium-sized stone with a long axis of about 10–20 cm, whose underside is not completely embedded in the mud or sand and that has some "interstitial" space above the bottom (a so-called "loose" stone), but is not situated on piled stones. This stone size will be the maximum that can be easily and carefully handled underwater, while harboring a considerable number of macrobenthos.

*b) Mesh size*

The selection of mesh-size for sieving bottom-samples is also taxa-dependent, and should be considered together with sampling area for respective taxa. In macrobenthos, some are to be roughly sorted out in the field by stirring samples in water, but others must be carefully sorted at the laboratory from samples that were fixed as soon as possible after collection. In the latter case, samples with bottom materials should be well sieved through a mesh before fixation, because the inclusion of large quantities of silt/mud particles in the samples will often result in the fixation failing. Considering the size of organisms as well as the ease of collection in the field, we recommend using a mesh size of 0.3–0.5 mm maximum diameter for samples smaller than 25 cm × 25 cm in area, and 0.5–1.0 mm mesh size for samples covering larger areas. Fine bottom particles which would otherwise, become obstacles for smooth sorting are thereby sufficiently removed. Once a standard mesh size for the investigation is selected, the same size (or finer) should be used for both the hand net and the mesh bag used for collecting and keeping samples underwater.

Where additional meiobenthic sampling will be included in the investigations, we recommend using another hand net with a smaller mesh size of 0.1–0.2 mm. The size (length and opening diameter) of this net might be slightly larger than the net collecting macrobenthos. If we put the latter net inside the former (meiobenthic net), we will simultaneously be able to receive both kinds of samples: of the macrobenthos, which will be retained in the (inner) net with the larger mesh size, and of the meiozoobenthos, which will pass through, the inner net and be collected in the external, meiobenthic net.

**Underwater handling**

This section describes the underwater sampling methods used by scuba divers. These methods are also applicable in waters less than 1 m using rubber boots or wet suits.

*a) Unconsolidated bottom without macrophyte vegetation*

Practically, the unconsolidated bottom can be divided into two categories:

1) soft bottom (sand, muddy sand, mud, etc.) and 2) bottom with coarse particles (gravel). In either bottom, we recommend using a hand net with a frame whose horizontal width is almost equal to the width of the quadrat for collecting macrobenthos together with bottom materials. The depth of the hand net should be large enough to contain all of the bottom materials within the quadrat.

At each sampling depth in respective observation sites representative types of substratum should be collected such as mud, sand, gravel, etc. Sampling should be replicated a minimum of three times for each type of substratum; more is preferable.

Where the bottom is soft enough (i.e., particles are sufficiently minute) to insert the frame of a hand net in to a sufficient depth (several centimetres from the surface) and to easily move it over the quadrat area, all of the soft-bottom materials within the quadrat can be scooped into the hand net together with macrobenthos, with the exception of highly movable animals such as fish and shrimp.

On gravel bottoms (where each particle is a few centimetres in diameter) the quadrat should be placed over an area where the gravel size is as uniform as possible. It is too difficult to insert a hand net into or move it through a gravel bottom. It is therefore recommended that each gravel particle be picked up from the bottom by hand and placed in the hand net carefully enough to prevent associated animals from escaping from the gravel. When all of the loose (i.e., not submerged) gravel particles are collected, the exposed bottom surface should be swept into the net by a rapid water current made by hand waving in order to collect macrobenthos together with smaller loose particles of the bottom.

After collecting samples of the bottom materials including organisms (hereinafter called bottom samples), the net should be tightly closed using rubber bands or "magic bands", especially when sampling by scuba diving. If the depth of the net is not sufficient to be closed or there are insufficient numbers of nets to take multiple samples at the same time, bottom samples collected by net can be carefully placed into a plastic bag of sufficient size during the underwater activities. Bottom samples within the net will be placed into sample bottles on board, and the net should be carefully washed for the next usage.

In order to record the particle size composition of bottom materials, a separate sample of all of the bottom materials themselves (hereinafter called bottom-material samples) is required, since the bottom samples collected for macrobenthos lose significant quantities of finer particles due to the large mesh size of the hand net. To collect bottom-material samples, the bottom surface is gently scooped with a small metal core or scoop. The captured bottom materials are then carefully placed into a plastic bag for later measurement.

*b) Macrophyte vegetation*

In many freshwaters macrophyte vegetation develops in the littoral zone where there is sufficient light for photosynthesis. The depth range of macrophyte vegetation may be determined by measuring the transparency of the water. The types of macrophyte community present will be influential on the associated macrobenthos community. Thus, representative macrophyte communities should be selected at each sampling depth in respective observation sites.

The macrophyte vegetation grows mostly on the unconsolidated (soft) bottom. Some samplers have been devised for collecting macrophytes on the soft bottom. These are generally designed to collect (cut out) macrophytes above the bottom (exposed in the water column) only, or to collect macrophytes together with the bottom materials. However, some macrobenthos may inhabit the leaves or stems of macrophytes, while other macrobenthos prefer the soft bottom under the vegetation. Thus, to evaluate macrophyte vegetation in the littoral zone, it is better to, separately but at the same time, collect both the macrophyte vegetation itself and the bottom materials below the vegetation. We will then be able to detect the utilization pattern of macrophytes by macrobenthos and the influence of the vegetation on macrobenthos dwelling on and in the bottom materials below the vegetation. We therefore recommend first collecting the macrophytes growing above the bottom (hereinafter called vegetation samples) and then the exposed bottom materials within the same quadrat area (hereinafter called bottom samples).

To collect vegetation samples, macrophytes can generally be easily collected by rooting them out by hand. This is preferable to cutting the stems by blade because if they are cut at the bottom surface, the roots remaining within the bottom materials may be obstacles to the smooth operation of the hand nets used to collect bottom samples. Therefore, we recommend carefully collecting samples of macrophytes together with associated macrobenthos by rooting them out.

When collecting vegetation samples, it is first necessary to establish a quadrat over a macrophyte community whose structure is as uniform as possible, to determine the species-specific patterns of macrophytes. Then the three-dimensional structure of the vegetation within the quadrat is observed, including relative abundance, maximum and average heights of respective species, and recorded using waterproof data sheets (or field notes), an underwater camera or video recorder.

After recording the physical structure of the vegetation within the quadrat, collect vegetation samples. Because macrophytes are often entangled with each other, it is better to roughly remove macrophytes growing outside of the quadrat before the sampling. Then the macrophytes inside the quadrat are collected; stems are to be gently drawn one by one, feeling for the roots in the soft bottom with fingers and gently pulled by hand until they are completely free. The macrophytes removed from the bottom with their roots are placed into a plastic bag or fine-mesh bag gently so as not to lose associated macrobenthos, especially gastropod snails that easily become detached from the macrophytes. After collecting all the macrophytes within the quadrat area, the opening of the bag is tightly closed. Special care is necessary because some macrophytes have slightly positive buoyancy when uprooted. Weight should be attached to the bag in order to achieve sufficient negative buoyancy when necessary.

When the collection of vegetation samples is finished, the bottom surface will be bare of vegetation. A bottom sample is then collected from this newly exposed (usually soft) bottom as described above for an unconsolidated bottom.

*c) Bedrock surface*

The surface of bedrock or large rocks on the bottom can be sampled using the same method as described for an unconsolidated bottom or macrophyte vegetation.

It is necessary to first find a bedrock surface with a uniform structure (uniform inclination, absence of cracks or holes), over which a quadrat can be established. After setting the quadrat on the bedrock surface, the area immediately outside of the quadrat should be roughly cleaned in order to remove macrobenthos which may become easily detached and contaminate the quadrat sample. Thereafter, a hand net should be placed on one side of the quadrat. Then, the bedrock surface within the quadrat is swept by water current made by hand or foot in order to blow the macrobenthos that are loosely attached to the surface into the net. It is usual for some animals to remain tightly attached to the bedrock surface. Thus, after collection of the loose animals with water current, the bedrock surface is cleaned with a brush or scraper to detach those animals that are tightly attached to the surface. The detached animals are then similarly collected in the hand net with water current.

## Stone unit sampling for loose stones

Loose (i.e., not submerged in the bottom) stones are highly heterogeneous microhabitats. The undersides of loose stones in particular are inhabited by a variety of unique animal species such as gastropods, amphipods and turbellarians. However, the quadrat sampling method used for soft bottoms cannot be used for sampling loose stones. We will therefore employ "stone unit sampling" for loose stones, wherein the sampling target is a selected loose stone, not a fixed area. Here we must select a moderate-sized stone whose long axis is 20 cm or less and can be easily handled.

At each stone unit sampling site it is ideal to have 15–20 replicates in order to achieve sufficient quantitative results.

For sampling, we recommend selecting a loose stone located on the soft bottom, rather than on piled stones. When a loose stone is selected, the conditions of the exposed surface of the stone should be recorded: underwater video or photographs, if available, should be taken from above and from the four lateral sides in order to record the number and species of large-sized animals on the stone surface. Before sampling the selected loose stone, the surface of the soft bottom surrounding the stone's horizontally projected area should be sufficiently cleaned to remove epibenthic animals, to avoid contamination of the stone sample. Then, the selected loose stone should be carefully placed into a plastic bag while keeping attached/associating animals on it (hereinafter called stone samples). The bottom exposed by the removal of the stone should be sampled immediately after the removal, using a hand net of the finest mesh size, and by creating a moderate water current by waving the hand to "wash" the fine particles and organisms off the bottom. The horizontal projection area or the entire surface area of the stone can provide an appropriate index for quantitative comparison.

## Treatment of samples in the field

Treatment of the above-mentioned samples may be different because of practical limitations in time, space, human resources, etc. When necessary, the collected samples can be kept fresh (alive) during transportation to the laboratory. However, researchers must ensure the absence of predatory species (e.g., insect larvae of Plecoptera, Odonata and Megaloptera, etc.) which directly influence the abundance of prey species within samples. Furthermore, since the field activities may be conducted in warm or even hot climates/seasons, special care is required to ensure that the samples are not overly heated. Prior preparation of cooling boxes is therefore recommended.

### a) Bottom samples

Since some fine bottom materials may absorb formalin during preservation, organisms have to be separated from the bottom materials before fixation. If conditions allow, rough separation of the organisms should be conducted on board immediately after collection. In order to separate organisms from bottom materials, we recommend the following procedures for treating bottom samples:

1) Prepare two or more empty buckets (at least one should be round, not square), a hand net with the minimum mesh size, a sample bottle, filtered water, and formalin (pure and 10 % solution).
2) Take (or pour) the bottom sample from the net or plastic bag and place it in a (round) bucket.
3) Add a sufficient amount of filtered water to the bottom sample (a few times as much in volume).
4) Strongly (but very carefully to avoid deformation of organisms) stir the bottom sample in the water by hand. Organisms (and other organic matter) with relatively low specific gravity will become adrift in the water.
5) Quickly pour the water with drifting organisms into a hand net with the minimum mesh size before they begin to re-settle (an empty bucket should be placed under the net in order to avoid accidentally losing the organisms remaining in the water).
6) Repeat steps 2–5 three or more times until most of the organisms, except for the heavy ones, have been separated from the bottom materials.
7) Pick up remaining heavy organisms such as thick-shelled bivalves and trichopterous larvae in the sand-case (which cannot be separated by the above treatments due to their mass) from the bottom materials, and place them in a bucket with filtered water.
8) Release the sample organisms filtered in the hand net into the bucket containing filtered water.
9) Replace the organisms collected in the bucket to a new sample bottle with water.
10) Pour formalin and water into the sample bottle to make a final concentration of about 10 % formalin.

If these processes cannot be conducted on board, the bottom samples should be temporarily fixed immediately after collection with formalin, which should be roughly separated from the fine bottom as soon as the samples arrive in the laboratory.

*b) Vegetation samples*

Rough separation of macrophytes and associated macrobenthos should be conducted for vegetation samples using the following process:
1) Prepare two or more empty buckets (at least one should be round, not square), a hand net with the minimum mesh size, a sample bottle or thick plastic bag (for collecting macrophytes), a sample bottle (for detached benthos), filtered water, and formalin (pure and 10 % solution).
2) Take the vegetation sample from the bag and place it into a (round) bucket.

3) Add a sufficient amount of filtered water to the vegetation sample (at least a few times as much in volume).
4) Carefully "wash" the macrophytes in the water by hand, thereby separating the loosely attached organisms from the macrophytes and setting them adrift in the water.
5) Quickly pour the water with drifting organisms into a hand net with the minimum mesh size to filter them out before they re-settle (an empty bucket should be placed under the net in order to avoid accidentally losing the organisms remaining in the water).
6) Repeat the above processes three or more times until the separation of organisms from the macrophytes is almost complete.
7) Release the organisms captured by the hand net into the bucket containing filtered water.
8) Replace the organisms collected in the bucket to a new sample bottle with water.
9) Pour formalin and water into the sample bottle to make a final concentration of about 10 % formalin.
10) Place the remaining macrophytes into another sample bottle or a thick plastic bag.
11) Add formalin solution to the macrophytes in the bottle or bag if conditions permit, and/or place the samples in cool conditions.

Since numerous macrobenthos (such as chironomid larvae) may still be tightly attached to the surface of macrophytes, the fixed macrophytes will be used not only for the analyses of macrophytes themselves but also for detailed examination of attached benthos in the laboratory.

*c) Stone samples*

Stone-unit sampling includes both stone-samples and under-stone samples. Under-stone samples should be treated in the same way as the bottom samples discussed above. When collected underwater, stone samples are placed in plastic bags with considerable quantities of water. While carrying these samples to the research vessel, some macrobenthos may become detached and drift in the water, while others may remain tightly attached to the stone surface. Stone samples should receive the following treatment:

1) Prepare a white plastic vat containing a few centimetres of filtered water, an empty round bucket, and an empty thick plastic bag.
2) Carefully remove the stone from the bag, and place it in the plastic vat.
3) Pour filtered water onto the stone to wash out the benthos loosely attached to the stone surface.

4) Place the washed stone in an empty plastic bag, add a proper amount of water or formalin solution, close the bag tightly, and put it in a cool place.
5) Pour the remaining water, from the original stone sample collection (containing drifting benthos) and the plastic vat, into the empty bucket.
6) Filter excessive water through a hand net or other appropriate mesh, then transfer the remaining water with benthos into a properly labelled sample bottle.
7) Add formalin to the sample bottle to achieve a 10 % formalin concentration.

The stone in the plastic bag should be treated with special care to avoid detaching any benthos until detailed treatment can be conducted in the laboratory.

*d) Sampling lots and labelling*
For the treatment and labelling of sampling lots, see sections 3.3.9 and 3.4.1.

## Laboratory methods
*a) Preliminary treatment of macrophytes*
Samples of macrophytes should be identified by species, and wet weights of each species should be measured in the laboratory. Organisms that were loosely attached to the macrophytes may remain in the bag in which the macrophyte samples were placed. These organisms are treated as attached to macrophytes. Then, the sample of each macrophyte species is placed in a bottle and is fixed with formalin solution for later examination along with the tightly attached organisms on the macrophytes.

*b) Examination of attached organisms on macrophytes*
Surfaces of the sampled macrophytes are carefully examined to collect attached organisms. The attached organisms are picked up with tweezers, and placed in different sample bottles for identification. These specimens are also preserved with formalin fixation. After examination, the macrophytes should be returned to the original bottle and preserved with formalin solution again.

*c) Sorting of macrobenthos according to major taxonomic groups*
Larger organisms are roughly separated from bottom materials and placed in formalin solution for preliminary fixation. Since formalin is highly toxic to humans, the fixed, roughly separated samples should be washed in water until the formalin is sufficiently diluted for further sorting by the following process:

1) Prepare a bucket (to stock wasted formalin), a vat (to contain the sample for sorting), a hand net, sample bottles, forceps, and binocular microscope.
2) Separate the fixed samples by filtering through a hand net of the minimum mesh size above an empty bucket (to receive wasted formalin solution).
3) Wash the fixed samples captured in the hand net with plenty of water to remove the majority of the formalin solution from the samples.
4) Release the washed samples from the hand net into a vat with a moderate amount of filtered water.

Organisms contained in the washed sample are then sorted into respective taxonomic groups with surgical forceps, by eye (for larger ones) and under a binocular microscope (for smaller ones). Fixed specimens are generally rather whitish and are thus more conspicuous against a black background.

Each taxonomic group should then be further identified, although the level of identification will differ between groups depending upon the status of the taxonomic information available for respective regions and the availability of highly-skilled taxonomists. Many of the mollusks and large crustaceans will be identified to the species or genus level. Many of the aquatic insect taxa such as Ephemeroptera, Plecoptera, Odonata, Megaloptera, Hemiptera, Coleoptera, and Trichoptera will also be identifiable at the species or genus level.

Finally, the number of individuals and wet weight for each taxon are recorded. Specimens should be weighed after removing effluent water on the surface body of samples by placing them on filter paper until wet spots vanish. Then, the weight is measured by electronic or torsion balance to the nearest 0.01g. From these values, we can obtain data on species composition and biomass in typical microhabitats of the respective freshwater bodies.

### 3.4.3 Meiobenthos
**Field methods**
*a) Collection under water*
Most of the meiozoobenthos reside in a 1–2 cm layer below the bottom surface. An area of ca. 20 cm$^2$ is sufficient for collecting the predominant species of meiozoobenthos. At each sampled depth for macrobenthos, a fixed area of the soft bottom (sand or mud/silt) should be carefully collected using a corer or scoop (e.g., a square metal corer with a 5 cm × 5 cm opening), and be placed directly into a plastic bag. Plastic bags with a zipper are convenient for ensuring tight closure. In selecting sites for collecting meiozoobenthos, special care should be taken to ensure there are no large benthos included in the sample.

*b) Sample treatment in the field*
Samples collected with bottom materials should be washed through a net with a mesh of ca. 0.3 mm in diameter to remove silty materials. In sandy bottoms, it is convenient to place the bottom materials in a Petri dish, add filtered water and carefully stir to separate the meiozoobenthos from the sand. Immediately after stirring, pour the upper layer of water into a bottle (volume ca. 100 ml). Repeat this procedure three times, to achieve a relatively complete extraction of the meiozoobenthos living amongst the sand particles. Larger animals that are visible and are not extracted by the above process should be taken from the sand using forceps. Bottled samples are fixed with a 2–4 % formalin solution.

If these procedures cannot be conducted in the field, they should be conducted immediately upon the samples' arrival in the laboratory.

*c) Sampling lots and labelling*
For the treatment and labelling of sampling lots, see sections 3.3.9 and 3.4.1.

**Laboratory methods**
Immediately after bringing a stone sample to the laboratory, the surface of the stone should be carefully examined, and the positions of attached benthos recorded. Preferably, each side of the stone (two horizontal and four lateral sides) will be photographed with a scale to evaluate the surface area and visually record the positions of larger benthos. Then the attached benthos are removed with a brush, scraper or forceps from each surface of the stone. The benthos removed from each side of the stone should be placed in separate sample bottles, and a formalin solution added for fixation and preservation.

Techniques for sampling and treatment of meiobenthos include many different methods that are highly specific to respective taxonomic groups. Meiobenthic animals are often seriously deformed after fixation by formalin and/or ethanol solutions, which makes further taxonomic identification impossible.

*In vivo* direct observations such as colour microphotographs, hand-made colour and/or black-and-white sketches and figures with scales, and other data therefore play a very important role in the treatment of "soft-body" groups such as Protista, Hydrozoa, microturbellarians, Gastrotricha, and some Rotifera and Oligochaeta. As a rule, these observations provide the possibility of identifying meiobenthic animals, at least, at the family, or even at the generic level. Microscopic nematods in the bottom sediments are better investigated with a dissecting microscope before treatment with formalin solution.

Gum-chloral liquid or Fore-Berlezet is the best universal medium for preparation of whole-mounts of meiobenthic invertebrates and is therefore recommended for the IBOY project. These can be easily prepared directly in the field, requiring only: a cover glass, microscope glass, gum-chloral liquid, pipet and/or forceps. Gum-chloral liquid is a gum arabic base permanent mounting media on a slide glass. (Hoyer's, mix the followings, 30–40 g gum arabic, 50 cc distilled water, 20 cc glycerin and 200 g chloral hydrate, at 60 °C and filtrate it through the glass-fiber filter. Rayne's, dissolve 10 g chloral hydrate in 10 ml distilled water, add 2.5 ml glycerin and 6 g gum arabic, mix avoiding bubbles, and wait 1 week before using [http://www.nmnh.si.edu/iz/copepod/techniques.htm]). This method provides excellent results for the investigation of biodiversity, taxonomic composition, and, even ecological data (stomach contents, for example) of meiobenthic (even some macrobenthic) animals, such as Rotifera, Acari and other small-sized Arthropoda, Gastrotricha, Nematoda, Plathelminthes, Oligochaeta, Tardigrada, etc.

1) Several Petri dishes with smoothly washed lacustrine sediments should be analyzed under a dissecting microscope.
2) Meiobenthic animals of different taxon should be concentrated on separate microscopic glasses (one per each morphological group), in a small droplet of fresh water.
3) Carefully remove as much water as possible from the glass by filtration paper after collecting several specimens of one group on one slide. Avoid drying out the animals that remain on the glass.
4) Add 1–2 drops of gum-chloral liquid directly to the concentrated animals, and carefully place the cover glass, avoiding air bubbles. If gum-chloral liquid under the cover glass is short, carefully add some droplets of gum-chloral liquid under the cover glass during the process.
5) The slide glass will be ready for preliminary microscopic investigation after 1–2 hours, although complete drying of whole-mounts will take several weeks.

Gum-chloral liquid strongly clarifies the internal structure of animals (especially their cuticular parts), which, in most cases is necessary for species-specific identification. If difficulties are encountered with taxonomic identification of the small meiobenthos, you may send colour photographs and/or drawings/sketches of their external features to one of the editors of this volume, who will be only too pleased to consult and provide advice on these matters.

### 3.4.4 Epi-microphyte (attached algae)
#### Field Methods
Attached algae are collected from the surface of stones. A 5 cm × 5 cm surface is necessary to collect attached algae from the stone. Stones with sufficient

surface area, therefore, should be collected concurrent with macrobenthos sampling. If stone-unit sampling stones are available, these can also be used to obtain samples of attached algae.

Use a small quadrat of 5 cm × 5 cm for sampling attached algae from the stone surface. The quadrat is placed on the flat, exposed surface of the stone. First, completely clean the stone surface outside of the quadrat, using a steel brush and water flush or completely cover it with vinyl-plastic. Then, carefully wash the area within the quadrat to completely detach any algae attached to the stone within the quadrat with a steel brush and water flush. The washing water is gathered in a bottle and diluted with distilled water. The volume of diluted water with detached attached-algae is measured, and is fixed with Lugol's solution at a final concentration of 3–5 %.

For the treatment and labelling of sampling lots, see sections 3.3.9 and 3.4.1.

## Laboratory methods

Attached algae are washed off of the attaching surface in the field, and then transferred to the laboratory. The sample should receive the following treatments in the laboratory.

1) Count algal cells under a microscope using a haemocytometer (e.g., square size = 300 μm × 300 μm, square numbers = 144, volume = 0.9 μl) for unicellular algae, and using a zooplankton counting chamber (e.g., square size = 1 mm × 1 mm, square numbers = 1000, volume = 1 ml) for filamentous algae, especially filamentous green algae such as *Cladophora*, *Oedogonium*, *Spirogyra* and *Ulothrix*.

2) Calculation of benthic algae densities.

Benthic algae (cells cm$^{-2}$) = N × (Ts/Cs) × (Cv/Tv)/A

 Where:

N: counted numbers (cells),
Ts: total numbers of square,
Cs: counted numbers of square,
Cv: chamber volume (ml),
Tv: total volume (ml),
A: area of quadrat (cm$^2$).

## 3.4.5 Planktonic organisms
### Sampling from the water column

We recommend that sampling points for planktonic organisms in the water column be established at bottom depths of 5 m, 10 m, and 20 m along the transect line of the littoral survey. We further recommend establishing additional sampling points, located on the offshore extension of the transect line, at arbitrary locations with sufficient depth to compare the littoral and offshore biodiversity. At each point, researchers on the research boat/vessel collect water samples using a water sampler (see below), from a set of water depths at 0–0.5 m (surface), 3–5 m, 7–10 m, and 15–20 m, plus 50 m and 100 m if applicable.

In the IBOY Project, researchers have to work with fixed samples. Identification at the species or genus level is expected to be very difficult for many of the taxa in most freshwater surveys partly due to deformation or even destruction of the body (cells) during fixation of samples and under-developed taxonomic study. Therefore, for bacteria, algal picoplankton, heterotrophic nanoflagellates, and ciliates, only their enumeration will be made, whereas nano- and micro-phytoplankton and mesozooplankton will receive identification at some taxonomic level (such as species, genus, and family) as well as their enumeration. Timoshkin et al. (1995) shows detailed sampling methods of pelagic animals and plants, with descriptive illustration useful for the identification of zooplankton and pelagic fishes of Baikal.

For the treatment and labelling of sampling lots, see sections 3.3.9 and 3.4.1.

## Collection and treatment of zooplankton

According to size, zooplankton can be categorized into several size-fraction groups: nanozooplankton (2–20 µm; mainly protozoan), microzooplankton (20–200 µm; protozoan, most rotifer species, young crustaceans), mesozooplankton (200–2000 µm; crustaceans, large rotifers), and macrozooplankton (centimeters long). Meso- and macro-zooplankton can be collected using plankton nets. We propose using a plankton net with an opening of 25–50 cm diameter (i.e., large enough to collect sufficient quantities of zooplankton) and a mesh size of 70 µm (i.e., fine enough to not miss any mesozooplankton). However, representatives of macrozooplankton such as the insect *Chaoborus* larvae, the opossum shrimp mysis, and, especially, the adult crustaceans are very active swimmers. Therefore, a standard plankton net (opening diameter of about 25–50 cm), which is very useful for mesozooplankton, can provide misleading information about the abundance of macrozooplankton organisms. More precise information about the abundance of macrozooplankton can be obtained by using a modified oceanic-type plankton net, which has an opening diameter of 0.8–1.0 m. The optimal mesh size of the net for macroplankton should be 300 µm and more, depending on the size spectrum of the macrozooplankton populations of the site being studied. The proportions of this larger net should be the same as for an ordinary plankton net.

In order to describe the vertical structure of a zooplankton community, samples should be collected from several water depth layers. We recommend taking vertical hauls with a lockable plankton net, with which we can collect zooplankton samples from a given water depth range.

When collecting zooplankton samples in a given water depth layer, the plankton net should be first placed at the deepest point of the layer, and then pulled vertically until it reaches the shallowest point. The opening of the net is then closed using a messenger, and the net is carefully drawn up into the boat.

After triple rinsing of a plankton net, the sample becomes ca. 250–300 ml in volume. The sample is poured from the net bucket into a plastic bottle of sufficient volume and is fixed with a 4 % formalin final solution. The samples are to be kept in a dark place. For prolonged storage, the bottle cap must be coated with paraffin or a wax/paraffin mixture.

Rotifers in this size range (20–200 μm) will pass through the plankton net. For more accurate sampling of zooplankton communities, we recommend using a Van-Dorn water sampler (see below).

After the samples in the bottle are fixed, several of the samples should be sub-divided into different bottles. Then, all of the organisms are counted for each preliminary identified taxa in each sample. An average number for each taxon is calculated and corrected for the whole sample volume. The number of each species in the sample volume serves as a basis for estimating their abundance in 1 $m^3$ (or, 1 litre) and under 1 $m^2$ of water surface. Biomass estimates can be obtained if the individual mass of each species is calculated and multiplied by its number.

## Collection and treatment of water sample, micro-, protozoo-, and phytoplankton

In order to measure chemical environmental variables and analyze the abundance of bacteria and phytoplankton, water samples are collected at the available depths at respective sampling points. Water samples should be collected using a clean Van-Dorn or a similar non-metallic water sampler with a volume of 3–10 litres (depending on the water quality and abundance of organisms).

After collecting water samples, a 1-litre portion of water sample should be poured into a plastic bottle, which should be kept in dark and cold conditions (10–20 °C degree). The remaining portion of the water sample is kept in a large plastic bottle in dark conditions.

For enumeration of bacteria, algal picoplankton (APP; 0.2–2 μm), heterotrophic nanoflagellates (HNF), pour a 100 ml portion of the water sample into a sampling bottle, and fix the sample using high quality glutaraldehyde solution with final concentration of 3–5 %. Glutaraldehyde (20–25 % solution) is appropriate for electron microscopy. Keep the fixed samples in dark and cold conditions (e.g., in a portable refrigerator) and place them in a laboratory refrigerator immediately upon completion of the field survey. Even with the samples stored in a refrigerator, the enumeration should be conducted within three weeks.

For enumeration of ciliates and the estimation of algal biomass (i.e., abundance of nanophytoplankton [2–20 µm] and microphytoplankton [20–200 µm]), portions of water samples should be divided and fixed with Lugol's solution. For enumeration of ciliates, a 500 ml portion of water samples is subdivided. The amount of water to be sub-divided for algal biomass estimation depends on the trophic condition of the freshwater; for example, 1000–2000 ml of water is used in the case of oligotrophic water, and 200–250 ml of water in meso- and eutrophic waters. The water samples either for ciliate enumeration or algal biomass estimation should be fixed with Lugol's solution at a final concentration of 1%. Keep the fixed samples in dark conditions until enumeration, which should be conducted within one month.

## Treatment of bacteria

1) Before filtering the water sample fixed with glutaraldehyde, prepare 50 µg/ml of DAPI (4'6-diamidino-2-phenylindole) solution (the DAPI solution) for staining. This solution should be filter-sterilized before use. (The DAPI solution can be stored in a refrigerator for several months.)
2) Rinse the filtration unit before use with filter-sterilized distilled water, and set up the filtration unit with a large-pore filter such as Whatman GF/C on a fritted glass base. Mount a black nuclepore filter of 0.2 µm pore size on to the other filter and assemble the funnel unit. The nuclepore filters should be moistened with filter-sterilized distilled water.
3) Filter an appropriate volume (0.1–1 ml from eutrophic water and >3 ml from oligotrophic water) of the water sample fixed with glutaraldehyde with a low (<0.3 ATM) vacuum. Release the vacuum immediately after filtration is completed.
4) Add 2 ml of DAPI solution (50 µg/ml) to the funnel and expose for 2–3 minutes of staining.
5) Put a drop of non-fluorescent immersion oil on a clean microscope slide.
6) After staining, draw the fluorescent dye through the filter with vacuum and rinse with 1–2 ml of filter-sterilized distilled water.
7) Remove the nuclepore filter, and mount it onto the drop of immersion oil.
8) Put one drop of immersion oil onto the center of the filter, cover with a cover glass and press slightly to flatten and expel the excess oil. Keep the microscopic specimens in the dark until examination.

9) Observe the microscopic specimens under ultra-violet excitation using an epifluorescent microscope in a darkened room. Use immersion oil at a magnification greater than x1000. Enumerate the bacterial cells with an eyepiece grid (ca. 40 × 40 μm). From 10–30 organisms per grid field may be appropriate for enumeration, and at least 300 cells for each sample should be counted.
10) Calculate bacterial densities:
    Bacterial cells ml$^{-1}$ = (C × N)/(F × V)
    where:
    C: conversion factor = (filtration area)/(area of the grid);
    N: counted number (cells);
    F: field number;
    V: filtered volume (ml).

## Treatment of algal picoplankton (APP)

1) Before filtering water samples fixed with glutaraldehyde, rinse the filtration unit with filter-sterilized distilled water, and set up the filtration unit with a large-pore filter such as Whatman GF/C on a fritted glass base. Mount a black nuclepore filter of 0.2 μm pore size on the other filter and assemble the funnel unit. The nuclepore filters should be moistened with filter-sterilized distilled water.
2) Filter an appropriate volume (ca. 1 to 2 ml) of the water sample fixed with glutaraldehyde with a low (<0.3 ATM) vacuum. Release the vacuum immediately after filtration is completed.
3) Put a drop of non-fluorescent immersion oil on a clean microscope slide.
4) Remove the nuclepore filter, and mount it onto the drop of immersion oil.
5) Put one drop of immersion oil onto the center of the filter, cover with a cover glass and press slightly to flatten and expel excess oil. Keep microscopic specimens in the dark until examination.
6) Observe microscopic specimens under green excitation using an epifluorescent microscope in a semidarkened room. Use immersion oil at a magnification greater than ×1000. Red fluorescence from APP cells will be detected under the green excitation.
7) Enumerate the APP cells with an eyepiece grid (ca. 40 × 40 μm) under green excitation. Although APP cell numbers are not as high as those of bacteria, at least 100 cells for each sample should be counted.
8) Calculate APP densities:
    APP cells ml$^{-1}$ = (C × N)/(F × V)
    where:

C: conversion factor = (filtration area)/(area of the grid);
N: counted number (cells);
F: field number;
V: filtered volume (ml).

## Treatment of heterotrophic nanoflagellates (HNF)

1) Before filtering the water sample fixed with glutaraldehyde, prepare 0.1 mol/l of Trizma-hydrochloride solution and adjust the solution at pH 4.0. Add primulin (fluorescent dye) at a concentration of 250 µg/ml (the primulin solution). Although you may see particles of primulin remaining in the solution, the particles will dissolve in the solution left in a refrigerator overnight. Primulin solution can be stored in a refrigerator for several weeks.
2) Filter an appropriate volume (2–5 ml from eutrophic water and >10 ml from oligotrophic water) of the fixed water sample with a low (<0.3 ATM) vacuum. Release the vacuum immediately after filtration is completed.
3) Add one ml of primulin solution to the funnel and expose for 3–5 minutes of staining.
4) Put a drop of non-fluorescent immersion oil on a clean microscope slide.
5) After staining, draw the fluorescent dye through the filter with vacuum and rinse with 3–5 ml of 0.1 mol/l Trizma-hydrochloride solution (pH 4.0).
6) Remove the nuclepore filter, and mount it onto the drop of immersion oil. Minimize the primulin fluorescence remaining in the filter, removing any excess by vacuum.
7) Put one drop of immersion oil onto the center of the filter, cover with a cover glass and press slightly to flatten and expel any excess oil. Keep microscopic specimens in the dark until examination.
8) Observe microscopic specimens under ultra-violet excitation using an epifluorescent microscope in a semi-darkened room. Use immersion oil at a magnification greater than ×1000.
9) Using an eyepiece grid (ca. 40 × 40 µm), enumerate nanoflagellates as HNF if they showed no obvious red chlorophyll fluorescence under green excitation. The enumeration should be conducted under ultra-violet excitation. Numbers of HNF cells in a field are not as high as those of bacteria, but as a minimum, more than 30 cells for each sample should be counted.
10) Calculate HNF densities:
    $$\text{HNF cells ml}^{-1} = (C \times N)/(F \times V)$$

where:
C: conversion factor = (filtration area)/(area of the grid);
N: counted number (cells); F: field number;
V: filtered volume (ml).

**Treatment of ciliates**
1) Concentrate a water sample fixed with Lugol's solution (ca. 500 ml) by natural sedimentation. Concentration in excess of ×1000 may be needed for samples from oligotrophic water, and greater than ×100 for those from eutrophic water.
2) Put one drop of the concentrated sample onto a haemocytometer, and enumerate ciliate cells under a microscope.

For identification of the dominant ciliate species, use non-fixed water samples, since the cell shapes of some ciliate species may change, or sometimes be destroyed, during fixation. This identification should be conducted within 1–2 hours of sample collection. A 10 % methylcellulose (15 cp) solution is useful for slowing down swimming ciliates. Patterson & Hedley (1992), Foissner & Berger (1996) and Foissner et al. (1999) may be helpful classification guides.

**Treatment of nano- and micro-phytoplankton**
1) Water samples fixed with Lugol's solution (200–2000 ml) should be kept for 48 hours for sedimentation of algal cells, and then concentrated to 0.5–1 ml.
2) Algal cells should be counted under a microscope using a haemocytometer.
3) Each algal cell number (cells/ml) should be estimated by: $C \times (N/n) \times (V/v)/S$

   where:
   C: individual counts (cell number),
   N: numbers of total square of haemocytometer,
   n: counting of square of haemocytometer,
   V: volume of concentrated sample (ml),
   v: volume of haemocytometer (ml),
   S: volume of sample (200–2000 ml).

# 3.5 Examples of observation sites

## 3.5.1 Clear lakes
In the present project, we propose including direct underwater investigation of

the littoral zone with the aid of scuba diving where conditions allow and it is practically available. Lakes Biwa and Baikal are representatives of clear lakes in the IBOY/DIWPA region, and provide a basis for comparing biodiversity observations in the littoral zone of other clear lakes.

For practical reasons, accessibility for researchers and availability of facilities are the most important factors for determining the research sites. We have provisionally selected the rocky shore along two capes (Tsuzurao-zaki and Kaizu-Oosaki) as research sites for Lake Biwa, and the rocky shore at Listvyanka and Bolshoye Koty for Lake Baikal. Rocky shores usually provide, in a minor topographical scale, a set of peninsulas (ridges) and bays (valleys) that continue under the water surface and largely determine the characteristics of the underwater landscape and substratum types. Erosional factors (lower deposition of SS (suspended substance), a thicker oxidized layer, and more coarse bottom particles) are more prevalent than depositional factors in peninsular areas, and vice-versa in bay areas. It is therefore ideal to establish research sites including both peninsular and bay areas.

In Lake Biwa, a depth of 21 m is always below the thermocline during the summer season (around 12–15 m in stratification seasons) and far deeper than the maximum depth of macrophyte growth (less than 10 m). All macrophytes and most mollusks, decapods, and littoral fishes are confined to the zone shallower than 10 m depth.

In Lake Baikal, the underwater landscape of the rocky areas is characterized by the abundant growth of branched green sponge, *Lubomirskia baicalensis*, forming a "forest" of sponge, usually situated between 5 m and 15–20 m deep. The densities of some unique species (e.g., the giant spiny amphipod [side-swimmer], *Acanthogammarus victorii* and of the gastropod [snail] *Liobaicalia stiedae*, etc.) increase deeper than 15 m.

At this time, we can conclude that the maximum safe diving depth of 21 m is sufficient to include the typical transitional patterns of biotopes along a bathymetric gradient. Thus, the survey methods detailed in this manual will be fully available for research in these lakes.

### 3.5.2 Turbid lakes

The lowland oxbow lakes in Borneo, such as lakes Tundai and Sabah, are representative bodies of turbid freshwater; as shallow, humic, and turbid freshwater they represent intermediate environments between lake and river. Due to a large seasonal fluctuation in rainfall, their maximum depths varied respectively between 5.5–8.9 m and 7.7–10.5 m in 1999 and 2000. The shorelines consist either of muddy bank or sandy slope with peat swamp vegetation surrounding the lakes. The peat swamp vegetation is submerged, forming swampy littoral zones. Due to its turbidity, the euphotic zone (the depth where there is sufficient light penetration to permit photosynthesis; where light intensity is greater than 1% of that at the lake's surface) was only 0.7–1.5 m.

With the exception of scuba diving, the survey methods mentioned in this manual are suitable for these turbid freshwaters.

The following four sites will be established at these oxbow lakes for biodiversity surveys (including running water environments):
1) the mouth of an oxbow lake (i.e., connection with the main river),
2) the central point of the arch-like shoreline of the lake,
3) the end point of the lake, and
4) connection points with small canals.

# References

Berg, L.S. 1965 Freshwater fishes of the U.S.S.R. and adjacent countries, vol. 3. Zool. Inst. Acad. Sco. U.S.S.R. Translated from Russian by I.P.S.T. Jerusalem.

Doi, A. 1997 A review of taxonomic studies of cypriniform fishes in Southeast Asia. *Japanese Journal of Ichthyology* 44: 1–33. (in Japanese)

Foissner, W. & Berger, H. 1996 A user-friendly guide to the ciliates (Protozoa, Ciliophora) commonly used by hydrobiologists as bioindicators in rivers, lakes, and waste waters, with notes on their ecology. *Freshwater Biology* 35(2): 375–482.

Foissner, W., Berger, H. & Schaumburg, J. 1999 *Identification and ecology of limnetic plankton ciliates*. Bayerisches Landesamt fur Wasserwirtschaft, Munchen.

Hubbs, C.L & Lagler, K.F. 1974 *Fishes of the Great Lakes Region*. Univ. Michigan Press, Ann Arbor.

Kajak, Z. 1971 Benthos of standing water. In: Edmondson, W.T. & Winberg, G.G. (eds.) *A manual on methods for the assessment of secondary productivity in fresh waters*. IBP Handbook no. 17. Blackwell Scientific Publications, Oxford, pp 25–65.

Kottelat, M., Whitten, A.J., Nartikasari, S.N. & Wirjoatmodjo, S. 1993 *Freshwater fishes of western Indonesia and Sulawesi*. Periplus Editions Ltd., Singapore.

Masuda, H., Amaoka, K., Araga, C., Ueno, T. & Yoshino, T. 1984 *The fishes of the Japanese archipelago*. Tokai Univ. Press, Tokyo. (in Japanese)

Nakabo, T. (ed.) 1993 *Fishes in Japan with pictorial keys to the species.* Tokai Univ. Press, Tokyo. (in Japanese)

Patterson, D.J. & Hedley, S. 1992 *Free-living freshwater protozoa: a colour guide.* Wolfe, London.

Riley, S.C. & Faush, K.D. 1992 Movement of brook trout (*Salvelinus fontinalis*) in four small subalpine streams in northern Colorado. *Ecology of Freshwater Fish* 1: 112–22.

Smirnov, V.V & Shumilov, I.P. 1974 *Baikal omuls.* Galaziy & Moskalenko (eds.). Nauka Publ., Novosibirsk. (in Russian)

Sideleva, V.G. 1982 *The lateral-line system and ecology of the Baikal sculpins* (Cottoidei). Novosibirsk. (in Russian)

Taliev, D.N. 1955 *Bychki-podkamenschiki Baikala (Cottoidei).* [Baikal Sculpins, Cottoidei]. USSR Acad. Sci. Publ. Moscow-Leningrad. (in Russian)

Timoshkin, O.A., Mazepova, G.F., Melnik, N.G., Obolkina, L.A., Tanichev, A.I., Bondarenko, N.A., Zemskaya, T.I., Kutikova, L.A., Pomazkova, G.I., Sideleva, V.G., Arov, I.V., Sheveleva, N.G., Afanasyeva, E.J., Mekhanikova, I.V., Shubenkov, S.G., Rusinok, O.T., Bekman, M.Yu., Logacheva, N.F., Alexsandrov, V.N., Podtyazhkina, M.M. & Pitulko, S.I. 1995. *Guide and Key To Pelagic Animals of Balkal With Ecological Notes.* NAUKA Publishers, Novosibirsk, 694 p. (in Russian with English in part)

Weber, M & de Beaufort, L.F. 1911 *The fishes of the Indo-Australian Archipelago. I. Index of the Ichthyological papers of P. Bleeker.* E. J. Brill, Leiden.

Weber, M & de Beaufort, L.F. 1913 *The fishes of the Indo-Australian Archipelago. II. Malacopterygii, Myctophoidea, Ostariophysi: I Siluroidea.* E. J. Brill, Leiden.

Weber, M & de Beaufort, L.F. 1916 *The fishes of the Indo-Australian Archipelago. III. Ostariophysi: I Cyprinoidea, Apodes, Synbranchi.* E. J. Brill, Leiden.

Weber, M & de Beaufort, L.F. 1922 *The fishes of the Indo-Australian Archipelago. IV. Heteromi, Solenichthys, Synentognathi, Percesoces, Labyrinthici, Microcyprini.* E. J. Brill, Leiden.

Weber, M & de Beaufort, L.F. 1929 *The fishes of the Indo-Australian Archipelago. V. Anacanthini, Allotriognathi, Heterostomata, Berycomorphii, Percomorphii: families Kuhliidae, Apogonidae, Plesiopidae, Pseudoplesiuopidae, Priacanthidae, Centropomidae.* E. J. Brill, Leiden.

Weber, M & de Beaufort, L.F. 1931 *The fishes of the Indo-Australian Archipelago. VI. Perciformes (continued).* E. J. Brill, Leiden.

Weber, M & de Beaufort, L.F. 1936 *The fishes of the Indo-Australian Archipelago. VII. Perciformes (continued) families: Chaetodontidae, Toxotidae, Monodacthylidae, Pempheridae, Kyphopsidae, Lutjanidae, Lobotidae, Sparidae, Nandidae, Sciaenidae, Malacanthidae, Cepolidae.* E. J. Brill, Leiden.

Weber, M & de Beaufort, L.F. 1940 *The fishes of the Indo-Australian Archipelago. VIII. Percomrphi (continued). Cirrhitoidae, Labriformes. Pomacentriformes.* E. J. Brill, Leiden.

Weber, M & de Beaufort, L.F. 1951 *The fishes of the Indo-Australian Archipelago. IX. Percomorphi (continued), Blennoidea.* E. J. Brill, Leiden.

Weber, M & de Beaufort, L.F. 1953 *The fishes of the Indo-Australian Archipelago. X. Gobioidea.* E. J. Brill, Leiden.

Weber, M & de Beaufort, L.F. 1962 *The fishes of the Indo-Australian Archipelago. XI. Scleroparei, Hypostomides, Pediculati, Plectognathi, Opisthomi, Discocephali, Xenopterygii.* E. J. Brill, Leiden.

# Drafting Team

Minako Ashiya, Lake Biwa Museum, Kusatsu, Japan
Olga I. Belykh, Limnological Institute SD RAS, Irkutsk, Russia
Nina A. Bondarenko, Limnological Institute SD RAS, Irkutsk, Russia
Atsushi Doi, Center for Ecological Research, Kyoto University, Otsu, Japan
Valentina A. Domysheva, Limnological Institute SD RAS, Irkutsk, Russia
Valentin V. Drucker, Limnological Institute SD RAS, Irkutsk, Russia
V. M. Gold, Krasnoyarsk State University, Krasnoyarsk, Russia
Nikolay G. Granin, Limnological Institute SD RAS, Irkutsk, Russia
Liba Z. Granina, Limnological Institute SD RAS, Irkutsk, Russia
Hiroki Haga, Lake Biwa Museum, Kusatsu, Japan
Dede I. Hartoto, Research and Development Centre for Limnology, Indonesia
Toshio Iwakuma, Hokkaido University, Sapporo, Japan
Lyudmila A. Izhboldina, Limnological Institute SD RAS, Irkutsk, Russia
Tamara V. Khodzher, Limnological Institute SD RAS, Irkutsk, Russia
Lyubov S. Kravtzova, Limnological Institute SD RAS, Irkutsk, Russia
Masayuki Kuwahara, Lake Biwa Museum, Kusatsu, Japan
Lev A. Levin, Institute for Biophysics SD RAS, Krasnoyarsk, Russia

Natalia G. Melnik, Limnological Institute SD RAS, Irkutsk, Russia
Igor B. Mizandrontsev, Limnological Institute SD RAS, Irkutsk, Russia
Hiroshi Morino, Ibaraki University, Mito, Japan
Hiroyuki Munehara, Hiroyuki Hokkaido University, Hakodate, Japan
Katsuki Nakai, Lake Biwa Museum, Kusatsu, Japan
Masami Nakanishi, Center for Ecological Research, Kyoto University, Otsu, Japan
Shin'ichi Nakano, Ehime University, Matsuyama, Japan
Kentaro Nozaki, University of Shiga Prefecture, Hikone, Japan
Lyubov A. Obolkina, Limnological Institute SD RAS, Irkutsk, Russia
Galina L. Okuneva, Irkutsk State University, Irkutsk, Russia
Valentina V. Parfenova, Limnological Institute SD RAS, Irkutsk, Russia
Galina V. Pomazkina, Limnological Institute SD RAS, Irkutsk, Russia
Galina I. Popovskaya, Limnological Institute SD RAS, Irkutsk, Russia
Natalia A. Rozhkova, Limnological Institute SD RAS, Irkutsk, Russia
Sergei V. Semovski, Limnological Institute SD RAS, Irkutsk, Russia
Pavel P. Sherstyankin, Limnological Institute SD RAS, Irkutsk, Russia
Michael N. Shimaraev, Limnological Institute SD RAS, Irkutsk, Russia
Valentina G. Sideleva, Zoological Institute RAS, St. Petersburg, Russia
Larisa P. Sorokovikova, Limnological Institute SD RAS, Irkutsk, Russia
Sulatri, Research and Development Centre for Limnology, Indonesia
Yasuhiro Takemon, Osaka Prefectural University, Sakai, Japan
Andrew I. Tanichev, Limnological Institute SD RAS, Irkutsk, Russia
Oleg A. Timoshkin, Limnological Institute SD RAS, Irkutsk, Russia
Takashi Toda, Lake Biwa Museum, Kusatsu, Japan
Masahide Yuma, Center for Ecological Research, Kyoto University, Otsu, Japan
Tamara I. Zemskaya, Limnological Institute SD RAS, Irkutsk, Russia

# Chapter 4: Latitudinal Biodiversity in Coastal Macrophyte Communities

*Chapter editors: Yoshihisa Shirayama, Ashley A. Rowden, Dennis P. Gordon & Hiroshi Kawai*

## 4.1 Introduction

The United Nations Environment Programme defines the coastal region as extending from upper tidal limits out across the continental shelf, slope, and rise (see Global Biodiversity Assessment, UNEP 1995). This definition includes rocky shores, sandy beaches, kelp forests, subtidal benthos, and the water column over the shelf, slope, and rise. Coastal systems are generally considered to encompass the Exclusive Economic Zones of nations, a strip approximately 200 nautical miles wide.

The importance of coastal ecosystems to humanity is vital as most of the world's people live within 80 km of the coast. Coastal ecosystems provide food and other resources, transportation, waste disposal, recreation, and inspiration. Some kelp forests, intertidal shores, and estuaries are among the most productive ecosystems in the world, and coastal fisheries are the richest in the world, with more than 75 % of the world's catch coming from coastal waters. Coastal ecosystems are also among those most heavily affected by humans, and threats to biodiversity are multiple and serious; they may also be synergistic. The effects of over-exploitation and pollution are increasingly obvious and serious (e.g. depletion or loss of food species, viral and bacterial diseases of marine organisms, contamination of food organisms, toxic-algal blooms), but the full consequences of alien species introductions, habitat modification or destruction, changes in UV-B radiation, and climate change have yet to be documented. Human pressure on the marine environment has never been so intense.

### 4.1.1 Importance of habitats

Article 7 of the Convention on Biological Diversity calls for the identification and monitoring of biodiversity, i.e. of ecosystems and habitats, species and communities, genomes and genes. In marine, as in terrestrial and freshwater ecosystems, it is well recognized that the biotic and physical attributes of habitats have a major influence on the diversity, distribution, and survival of organisms. Changes in the nature of marine habitats can cause rapid changes in biodiversity composition, including species of commercial interest. For example, seagrass beds in estuarine and open-coast environments influence local species diversity including fish species whose juveniles use such beds as nursery areas. In the tropics, scleractinian and soft corals structure habitats three-dimensionally, locally increasing biodiversity by providing spatial niches for a wide range of invertebrates, vertebrates, and algae, in turn influencing food-web structure and increasing the complexity of biological interactions. In shallow temperate waters, large seaweeds, bryozoans, hydroids, and tubeworms play a similar role, and in deeper water, as on seamounts, tree-like black corals, gorgonians, scleractinians, and stylasterid hydrocorals are important.

Alterations to natural habitats may be caused by natural processes or human activities. The latter may be direct (e.g. input of terrestrial sediments from forest clearance, pollutants, mariculture, benthic trawling, dumping of offal from fish-factory ships, introduction of alien species including disease organisms) or indirect (e.g. climate change). Changes in population abundance or density, or removal of species (especially keystone, trophically important, or habitat-structuring species), can initiate a cascade of effects that may fundamentally alter biodiversity. Destructive fishing techniques seriously affect marine communities structured by slow-growing coralline and tree-like organisms on rocky bottoms, but very few impact and monitoring studies have been carried out in such critically important subtidal habitats.

### 4.1.2 Importance of monitoring

Inventory and monitoring of biodiversity are crucial for identifying or clarifying the pressures that impact on ecosystems, the rates at which those pressures are operating, present and likely states of those ecosystems, and the actions or responses needed to mitigate or stop negative pressures. The pressure-state-response model is among the more helpful models being used to guide the process of asking the right questions and formulating monitoring programmes.

Monitoring generally requires repeated sampling over time. Effective monitoring requires that sampling is replicated to detect variations over short to long time periods, and at more than one location. This means that sampling design is a very important part of devising a monitoring strategy. Studies of distribution and abundance generally require sampling through time to detect patterns that could potentially change over time-scales of days (state of tide, fluctuations in light, temperature, and atmospheric pressure), seasons, years, and decades. Sampling frequency must therefore coincide with whatever variable is being measured. For example, if one is sampling every month, additional sampling should be done within months to demonstrate that variations within a day, between days, and between weeks is less than the variation found over months, seasons, and years. Ideally this procedure should be repeated at more than one location.

### 4.1.3 Baseline studies

Baseline studies refer to data that are collected to define the present state of a habitat, population, or biodiversity in general, in relation to physical parameters and anticipated impacts. Before conducting a baseline study it is important to ask some initial questions. What is being measured? What changes could be anticipated and why? What spatial and temporal scales are appropriate? One-off baseline studies are generally of limited value if they are not replicated in time and space. They have very little predictive power.

Baseline data usually include:
1 The presence and/or abundance of species or other units;
2 Other dependent data (e.g. size and distribution of rock pools, boulders, caves, canopy species, and other features of habitats affecting marine occurrences).
3 Appropriate influential abiotic variables (see below).
4 Human variables.

As the goals and scales of inventorying and monitoring programmes may change with time the baseline data collected should be sufficiently robust to accommodate such changes. Provided the data represent a robust sample of the system under study, baseline data can be used to calibrate methods of Rapid Biodiversity Assessment (see below).

## 4.2 Goals for Monitoring Coastal Ecosystems

For the purposes of IBOY, we suggest monitoring three codependent gradients in the coastal zone throughout the DIWPA region to a depth of 10 m (15 and 20 m are optional). These are latitudinal and related gradients or clines; gradients induced by human impacts; and temporal gradients (long-term monitoring).

## 4.2.1 Latitudinal and related gradients

Gradients of distribution of organisms have been identified in the sea. The details of some of these gradients are still being clarified, however, as they are not all necessarily straightforward. For example, in the northern hemisphere there is said to be a latitudinal increase in the numbers of species from the Arctic to the tropics. This is not the case in the southern hemisphere, however, where some of the highest diversities for soft-sediment biota have been found at 38°S off the Victorian coast of Australia, and Antarctica has high diversities for many taxa. It is also still not clear how diversity changes in soft sediments from the continental shelf into the deep sea, as there are relatively few data for many groups of organisms and the deep sea is badly under-sampled. One of the best-known diversity patterns is that of regional-scale decreases in coral genera from the Malaysian archipelago eastwards across the Pacific Ocean and westwards across the Indian Ocean, with the lowest diversity in the Caribbean. Similar patterns have been found for mangroves and gastropod snails. On a smaller scale, some South Pacific islands have an E–W rainfall gradient that may be expected to have local effects in lagoonal and littoral environments.

Inventory and monitoring of biodiversity in the coastal environment (as defined above) are necessary to clarify the details of such gradients and how they may shift as a consequence of natural and anthropogenic perturbations.

## 4.2.2 Long-term monitoring

DIWPA monitoring sites have the potential to become Long-Term Ecological Research (LTER) sites.

# 4.3 Site Selection

## 4.3.1 Regional level

Within the DIWPA region, from the Russian subarctic through the tropics to New Zealand's subantarctic islands, there is a huge range of coastal marine ecosystems and habitats. As a preliminary to selecting biodiversity monitoring sites in the DIWPA region, those countries that have not yet devised coastal classification schemes would benefit from doing so. Various schemes have been devised, including one for the marine realm globally. That is a hierarchical scheme that begins "coarse-grained" (zoogeographic realm): then proceeds through "medium-grained" (the finest level possible at a regional scale) to "fine-grained" (national and provincial scale), at which point it becomes a genetic classification, subdividing coastal environments, offshore environments, pelagic environments, coast-associated habitats, living reefs, and critical habitats into finer categories (see appendix). Use of a consistent classification scheme, like that above, throughout the DIWPA region would more easily allow selection and subsequent monitoring of comparable sites (e.g. shallow subtidal kelp beds in Japan with those in New Zealand; hermatypic coral reefs in the northern Ryukyu Islands with those in the southern Great Barrier Reef).

Coastal ecosystems represent the margins of larger ecosystems, varying considerably depending on atmospheric, oceanographic, geological, and historical factors. Accordingly, forty nine Large Marine Ecosystems have been delineated globally, representing regions of ocean space from deltas and estuaries to the seaward boundaries of continental shelves and coastal current systems. They are regions characterised by distinct bathymetry, hydrography, productivity, and tropically linked populations. Those in the DIWPA region comprise: West Bering Sea, Sea of Okhotsk, Oyashio Current, Sea of Japan, Kuroshio Current, Yellow Sea, East China Sea, South China Sea, Sulu-Celebes Seas, Indonesian Seas, Northern Australian Shelf, Great Barrier Reef, New Zealand Shelf, and Insular Pacific.

The Global 200 programme provides a similar schema of delineation, organizing the world's most outstanding ecoregions (233 identified) biogeographically by habitat type within terrestrial, freshwater, and marine realms (Olson & Dinerstein 1998). Of these, 18 are located within the DIWPA region. Ideally, it is desirable to locate biodiversity monitoring sites in each of the Large Marine Ecosystems/Global 200 Ecoregions in the overall DIWPA region. As Brunckhorst and Bridgewater (1994) have pointed out, bioregions should be the ultimate management units for sustainable societies, affecting consequent planning and management purposes, in which ecologically sustainable use becomes the management paradigm.

More practically, in order to compare biodiversity on a global scale, the IBOY study requires at least three study sites in each 20° bin along the proposed latitudinal transect between 50°N and 50°S.

## 4.3.2 Local selection criteria

A two-tiered approach to biodiversity monitoring is recommended, utilizing core and satellite sites. Intensive baseline studies and monitoring will be carried out at core sites using all standard methodologies; at satellite sites, only some methodologies need be employed or data collected. Core and satellite sites may be selected on the basis of the following criteria.

## Infrastructure

Long-term monitoring (over years to decades) is most easily accomplished in proximity to a research facility (e.g. a marine laboratory) where there is likely to be accommodation and ongoing research programmes. Automatic 24-h monitoring of physical data is possible when remote instrumentation is connected directly to a computer in a laboratory (see section 5 on methodology). A major benefit of locating monitoring sites near a research facility is that routine measurements of biodiversity and physical variables can often be carried out relatively cheaply using student labor or other on-site/near-site human resources. It also means that a commitment to long-term monitoring is more easily achievable. Marine station networks may facilitate planning and coordination of research efforts.

## Baseline information

For a variety of historical, geographic, resource, and other reasons, some areas of coastline are better known biologically and physically than others. The existence of historical data allows closer comparisons between former and current states, and may help in the process of site selection when potential monitoring sites are otherwise closely similar. In addition such information would be useful for future compilation of biological information.

## Reasonably natural environment

(Pristiness according to MARS definition). A goal of the regional monitoring programme is within-region comparisons of biodiversity and biotic change. It is therefore desirable that monitoring should be carried out in areas that are as natural as possible. It would be advantageous to locate monitoring programmes within marine protected areas (MPAs), for example. There are a variety of marine reserves and usages throughout the DIWPA region, ranging from controlled exploitation of certain species (usually line-fishing of reef fish) to complete no-take zones. The latter type of MPA is not common but could be ideal for monitoring activities provided other criteria are satisfied.

## Long-term stability of the site

It needs to be ascertained if a proposed monitoring site is likely to remain the same during the monitoring period. Thus it may be necessary to determine if coastal development or modification of an adjacent catchment is intended. It is important to eliminate human-caused variables as far as possible.

## Accessibility

Sites that are more natural in biological character, i.e. containing ecosystems or habitats that are unmodified or scarcely modified by human activities, are frequently the most remote and difficult to access. Some coasts are also subject to greater wave exposure and are less able to be regularly sampled. Deeper-water habitats are expensive to sample and monitor, and successful occupation of the same station for extended periods or over the long term is dependent on sea-surface state.

## Biological character

Pre-selection criteria can include known biodiversity values; is the candidate site biodiversity-rich? Is it representative of a wider biotic ecosystem or realm? Is there a significant number of rare species, etc.? It is also important that the target habitats, i.e. 'homogeneous' macroalgae-hard and/or seagrass-soft substratum habitats with a shoreline extent of 20–200 m should be available in the site.

### 4.3.3 Application of selection criteria

Potential biodiversity monitoring sites can be rated according to each criterion (excellent, reasonable, poor, no data) and ranked according to their scores.

### 4.3.4 Potential availability of no-fishing/no-take areas for stability of long-term monitoring

As discussed in 4.3.2 above, monitoring can be effectively carried out in protected areas. Ideally, these should be completely no-take marine reserves in which no extraction of organisms takes place. Unfortunately, such reserves are rare anywhere in the world; these should be established as a matter of principle. However, in most maritime countries there are already many areas of seafloor that have been declared no-fishing and/or no-entry areas for sectoral reasons – because of their restricted nature these areas constitute de facto reserves (e.g. military areas and cultural sites). All of these constitute areas where potentially undisturbed monitoring can take place under appropriate circumstances.

### 4.3.5 Marine BioRap – Identifying biodiversity priority areas

Marine BioRap is a methodology and set of analytical tools developed in Australia for identifying and assessing, in less than 18 months, priority areas of marine biodiversity. It is a decision support tool that can help planning and decision-making by identifying priority areas from local to ocean-basin scales.

BioRap also uses biodiversity itself (or surrogates of biodiversity) to identify priority areas, while taking into consideration other factors, and precedes using iterative approaches. BioRap is an approach that can be used in selecting among candidate monitoring sites when there are a number of similar sites to choose from.

## 4.4 Sampling Protocol

### 4.4.1 Sampling strategy

At each study site a stratified random sampling strategy will be employed, with strata representing vertical heights above and below the low water datum. That is for each study habitat, five random replicate samples are to be taken at high, mid and low intertidal positions and 1, 5 and 10 m subtidal water depths (15 and 20 m depth strata are optional). The most expedient randomization procedure should be adopted. The sampling programme at each study site should take place at least once a year, during the period of expected highest diversity, and commence by the end of 2002 if possible.

### 4.4.2 Sampling methodology

The sampling methodology hereafter described is a minimum requirement to be done at each site for IBOY activity. Ideally, there are many factors to be measured and subjects to be studied. All of these are described later as recommendations for more extensive studies (see 4.4.4).

At each random replicate sample location both non-destructive and destructive sampling will be undertaken according to the following protocol.

**In-situ observation (non-destructive)**

A photographic image record (digital or film) should be made immediately prior to sampling. If conditions (e.g. poor visibility) do not permit a photographic record to be made, a hand-drawn map should be constructed as an alternative. All macrophytes and conspicuous macrofauna (> 2 cm length) within a quadrat sample will be identified in-situ, and either counted or a percentage of cover estimated using a standard technique. For macroalgae-hard substrate habitats a $1 \times 1$ m quadrat will be utilized, whilst for seagrass-soft substrate habitats a $50 \times 50$ cm quadrat will be sampled. Counts will be made of solitary fauna, erect colonial organisms and seagrass plants. Percent cover estimates (using a standard technique) will be made for canopy and understorey macroalgae, and encrusting colonial organisms.

## Direct removal (destructive)

A photographic image record (digital or film) should be made immediately prior to sampling. All macrophytes and fauna within a quadrat or core sample will be carefully and completely removed. For macroalgae-hard substrate habitats a 25 × 25 cm quadrat will be sampled, whilst for seagrass-soft substrate habitats a 15 cm diameter cylindrical core (to 10 cm substrate depth) will be utilized. Both quadrat and core shall have a 63 μm mesh bag attached, into which macrophytes and fauna should be collected without significant loss of material. Hand scrapers will be used in macroalgae-hard substrate habitats in order to facilitate removal of attached organisms.

In the first year of sampling, the 25 × 25 cm quadrat utilized for directly sampling the macroalgae-hard substrate shall form a sub-sample (always the same position within the larger sample) of a 50 × 50 cm quadrat, from which only macroalgae shall be completely removed. This latter sample is taken in order to ensure sufficient algal reference material to support the in-situ observation.

The surface and bottom seawater temperature should be measured at each sample location. In addition, the substratum should be visually classified according to the standard Wentworth convention for the description of sediments.

### 4.4.3 Initial processing of direct samples

Resulting samples should be sieved on nested meshes of 1 mm and 63 μm. Macrophytes remaining on the 1 mm sieve should be carefully washed (and if necessary scraped) over the mesh to remove associated macrofauna. Both the floral and faunal component of the 1 mm sample are to be retained, but should be stored separately. The material retained on the 63 μm sieve will largely comprise meiofauna. All three portions of the sample should be separately fixed and preserved using 5 % neutralized formalin* (2 % formaldehyde) in seawater.

### 4.4.4 Recommendations

The above protocol constitutes the minimum standardized sampling requirement for the proposed biodiversity determination, comparison and monitoring study (IBOY). The following recommendations represent actions which are considered to be useful optional additions to the programme: (1) Sampling more frequently than once a year, e.g. during potentially separate

---

\* concentrated formalin (= 35 % formaldehyde) saturated with borax (sodium hexaborate)

periods of highest diversity for macrophytes and associated fauna. (2) Sampling of additional habitats that occur at a selected study site, e.g. mangrove, coral reef, unvegetated sediment. (3) Creation of a macrophyte and macrofauna reference collection for the study site. (4) Taking of additional samples for future molecular studies (fixed and preserved in 100 % ethanol). (5) Compilation of a site species inventory from existing information. (6) Construction of site history, e.g. adjacent terrestrial land 'use', potential anthropogenic impacts.

## 4.5 Subjects To be Studied and Monitored

A regional approach to monitoring coastal biodiversity invites the question, what aspects of biodiversity may be monitored that can be compared throughout the region? Four subjects are recommended here for study and monitoring at core sites:
- species inventory of selected taxonomic groups
- abiotic and biotic parameters
- habitat mapping
- all-biota taxonomic inventory.

The minimal requirements outlined above for sampling will provide samples that fulfil most of the subjects discussed below. However, it is not possible to carry out all of the subjects listed below for each participating sampling site due to lack of funds, facilities and human resources. Strategies to overcome these problems will be discussed later.

### 4.5.1 Species inventory of selected taxonomic groups

Major taxa to be studied may be selected by a variety of criteria including representation throughout the DIWPA region, ease of identification by non-experts, commonness, ecological role (keystone species, habitat-structuring, trophic importance), use as an environmental indicator, etc. Selected species from the following groups are recommended:
- macroalgae
- seagrasses
- mollusks
- decapod crustaceans
- echinoderms
- fish
- cnidarian corals.

Depending on locality and geographic area, optional taxa can include selected species of:
- sponges
- other macro-invertebrates (large bryozoans, hydroids, ascidians)
- marine reptiles
- seabirds
- marine mammals.

## 4.5.2 Abiotic and biotic parameters

Easily measurable physical and biological parameters influencing or associated with coastal biodiversity include:
- temperature
- salinity
- water chemistry (C, H, N, O, nutrients, etc.)
- pH
- suspended sediments
- currents
- light
- chlorophyll a.

Although the sampling protocol only requests temperature measurement, it is recommended that the above listed parameters be measured from the sea surface down to 20 m depth. To ensure data quality and to facilitate regional comparisons, continuous observation by multiple-sensor data-loggers is highly desirable. Standardized methodology may be possible through the mass production of sensing apparatuses.

## 4.5.3 Habitat and biodiversity mapping

As mentioned in the sampling section (4.4.4), it is necessary to find a homogeneous macroalgae-hard and/or seagrass-soft substratum habitat in each site. Information obtained from habitat mapping will provide data necessary for selecting the sampling site at each location.

Mapping can be a two-tiered exercise. At one level, entire coastlines can be mapped biologically, based on a variety of data sources, though it is not mandatory for each participating site. If such maps already exist, as they do for parts of some DIWPA countries, again they can facilitate the selection of biodiversity monitoring sites. At a finer level, detailed maps may exist for some marine protected areas, and should be carried out in areas selected for monitoring. If maps of coastlines do not already exist, then the production of habitat maps at monitoring sites can contribute to the later production of larger-scale coastal maps.

Coastal-zone maps at a variety of scales may already exist for mangroves and coral reefs. Maps can also depict the distribution of macroalgae, subtidal biogenic structures (e.g. bryozoan mounds, tubeworm reefs, sponge beds), shellfish beds, seagrasses, seabird and turtle nesting sites, and hauling grounds for pinnipeds. Use of GIS (Geography Information System) can overlay and correlate associated sediment, hydrographic, and other data obtained from on-site and remote (aerial and satellite photography, sonograph) measurements.

### 4.5.4 Species inventory and sampling

Coddington et al. (1991) have provided strategies for species inventories, including:
- Use proven collection methods for different taxonomic groups in order to standardize techniques with previous and future researchers.
- Keep the number of collection methods for each group to the minimum necessary, but maximize the independence among methods.
- Use general protocols that work in plot-based or plotless sampling.
- Keep the sampling unit general, simple, and comparable: time spent sampling is perhaps the best unit of measure. Sample units should be small enough to permit among-sample comparisons.
- Large samples should be reassembled from smaller replicate samples.
- Data collected should permit variation to be estimated and analysed, especially with respect to site, season, sampling method, etc.
- Samples of species and individuals per species should be sufficient to construct species abundance distributions that can be used to estimate species diversity.
- Since some sampling methods tend to under-record rare taxa, sampling should be designed with the aim of reliably reproducing the population characteristics (as distinct from sampling-error effects).
- Voucher specimens of each species must be conserved to ensure taxonomic consistency and accuracy of identification.

More detailed information can be found in Global Biodiversity Assessment (Heywood 1995: p. 478).

The sampling protocol described in section 4 was designed to fulfil all of these criteria.

### 4.5.5 All-biota taxonomic inventory

Where appropriate, some core monitoring sites throughout the region in similar habitats should be chosen for all-biota taxonomic inventory (ABTI). These could be considered as core sites for long-term monitoring beyond the immediate scope of the IBOY project.

Impediments to an ABTI include the availability of systematic expertise in the short and long term and funds for capacity building. It is recommended that, where possible, the same taxonomic experts be available for shared comparative inventory across the DIWPA region. The availability of expertise will determine whether an inventory of target taxa will be intensive or whether some form of rapid assessment will be used. The latter approach can be effective if it allows for repeatability in the discrimination of recognizable but unnamed taxa (so-called RTUs).

## 4.6 Strategies for future activities

### 4.6.1 Sampling kit
To ensure the highest degree of standardization practicably possible it is desirable to seek central funding for the provision of sieves and digital camera equipment (part of minimal sampling kit).

### 4.6.2 Future activities
In the near future, it is proposed that a database containing contact addresses/emails of the study participants and the details of all selected study sites will be constructed. Study site details (e.g. precise latitude/longitude, habitat characteristics, etc.) have been solicited by questionnaire. Information pertaining to the study – its aim, sampling protocol, map of study sites, list of participants, etc. – will be posted on a soon to be developed DIWPA webpage (with support from CoML). It is essential that all study participants communicate their sampling schedule directly by means of the group email list and via the webpage.

In order to analyse the initial results of the study (data for macrophytes and conspicuous macrofauna), a workshop will be organized for all participants at the end of 2002 or the beginning of 2003. Currently there is no precise agreement as to the mechanism by which the samples of fauna (macrofauna and meiofauna) not examined in-situ will be identified, and the results compiled and analysed. However, one possibility is to assemble a team of 'itinerant' post-doctoral researchers who can be collectively responsible for ensuring that the biodiversity assessment of each study site is completed.

It is envisaged that collaboration will be established and maintained with related projects within programmes such as BIOMARE.

# References

Brunckhorst, D.J. & Bridgewater, P.B. 1994 A novel approach to identify and select core reserve areas, and to apply UNESCO Biosphere Reserve principles to the coastal marine realm. In: Brunckhorst, D.J. (ed.) *Marine protected areas and biosphere reserves: Towards a new paradigm.* Australian Nature Conservation Agency, Canberra, pp.12–17.

Kingsford, M. Battershill, C. 1998 *Studying temperate marine environments: A handbook for ecologists.* Canterbury University Press, Christchurch.

Jeffrey, S.W. Mantoura, R.F.C. & Wright, S.W. 1997 *Phytoplankton pigments in oceanography: guidelines to modern methods.* Monographs on Oceanographic Methodology 10.

NOAA, 1997 *Remote sensing for coastal resource managers: An overview.* US Department of Commerce, Washington, D.C.

Olson, D.M. & Dinerstein, E. 1998 The Global 200: A representation approach to conserving the Earth's most biological valuable ecoregions. *Conservation Biology* 12: 502–15.

Ormond, R.F.G., Gage, J.D. & Angel, M.V. (eds.) *1997 Marine biodiversity: patterns and processes.* Cambridge University Press, Cambridge.

Phillips, R.C. & McRoy, C.P. (eds.) *1990 Seagrass research methods.* Monographs on Oceanographic Methodology 9.

Ray, G.C. 1977 A preliminary classification of coastal and marine environments. *Bulletin of the Marine Park Research Stations* 1: 123–7.

Ray, G.C. 1988 Ecological diversity in coastal zones and oceans. In: Wilson, E.O. & Peter, F.M. (eds.) *Biodiversity.* National Academy Press, Washington, D.C., pp.36–50.

Sherman, K. & Alexander, L.M. (eds.) 1986 *Variability and management of Large Marine Ecosystems.* AAAS Selected Symposium 99. Westview Press, Boulder.

Sherman, K., Alexander, L.M. & Gold, B.D. (eds.) 1990 *Large Marine Ecosystems: Patterns, processes and yields.* AAAS Press, Washington, D.C.

Snedaker, S.C. & Snedaker, J.G. 1984: *The mangrove ecosystem: research methods.* Monographs on Oceanographic Methodology 8.

Sournia, A. (ed.) 1978 *Phytoplankton manual.* Monographs on Oceanographic Methodology 6.

Stoddart, D.R. & Johannes, R.E. 1978 *Coral reefs: research methods.* Monographs on Oceanographic Methodology 5.

Tranter, D.J. & Fraser, J.H. (eds.) 1968 *Zooplankton sampling.* Monographs on Oceanographic Methodology 2.

UNESCO 1984 *Comparing coral reef survey methods*. UNESCO Reports on Marine Science 21.

Von Alt, C., DeLuca, S.M., Glenn, J.F., Grassle, J.F. & Haidvogel, D.B. 1997 *LEO-15: monitoring and managing coastal resources*. Sea Technology 38: 10–16.

Ward, T.J., Kenchington, R.A., Faith, D.P. & Margules, C.R. 1998 *Marine BioRap guidelines: Rapid assessment of marine biological diversity*. CSIRO, Perth.

# Appendix 1: Global 200 marine ecoregions occurring within the DIWPA region

## Marine ecoregions:
### Large deltas, mangroves, and estuaries
*Indomalayan:*
185. Mekong River Delta mangroves: Vietnam, Cambodia
187. Sundaland and eastern Indonesian archipelago mangroves: Indonesia

*Australasian:*
189. New Guinea mangroves: Papua New Guinea, Indonesia

### Coral reef and associated marine ecosystems
*Western Pacific Ocean:*
201. Isthmus of Kra marine ecosystems: Thailand, Malaysia
202. Nansei Shoto marine ecosystems: Japan
203. Sulu Sea: Philippines, Malaysia
204. Sulawesi Sea: Philippines, Indonesia, Malaysia
205. Banda-Flores Seas marine ecosystems: Indonesia
206. Northern New Guinea and Coral Sea marine ecosystems: Papua New Guinea, Indonesia, Solomon Islands
207. Micronesian marine ecosystems: Palau, Federated States of Micronesia

*Southern Pacific Ocean:*
209. South Pacific marine ecosystems: Vanuatu, Fiji, New Caledonia, Samoa, Tonga, Tuvalu
210. Great Barrier Reef: Australia
212. Lord Howe Island and Norfolk Island marine ecosystems: Australia

**Coastal marine ecosystems**
*Western Pacific Ocean:*
222. Yellow Sea and East China Sea: China, North Korea, South Korea, Japan

*Southern Pacific Ocean:*
228. South temperate Australian marine ecosystems: Australia

**Polar and subpolar marine ecosystems**
*Antarctic seas:*
230. New Zealand marine ecosystems: New Zealand

*Arctic Ocean and seas:*
231. Bering and Beaufort Seas: Russia, USA, Canada
232. Sea of Okhotsk and northern Sea of Japan: Russia, Japan

# Appendix 2: Questionare form and invitational letter to be distributed at potential participating sites*

## Invitation to the International Biodiversity Observation Year (IBOY) of DIVERSITAS International of the Western Pacific Asia (DIWPA) Coastal Module in cooperation with Census of Marine Life (CoML) and Japan Society for the Promotion of Science (JSPS)

*Yoshihisa Shirayama Seto Marine Biological Laboratory, Kyoto University, Shirahama-cho, Japan*

There is growing global interest in the conservation of biodiversity. DIVERSITAS is a program established under the support of UNESCO, SCOPE, IUBS and others. Its aim is to promote and facilitate international research activities for the analyses of biodiversity. DIVERSITAS has designated 2001 to 2002 as an International Biodiversity Observation Year (IBOY), and certain projects to study biodiversity on a global scale are ready to start in this year.

Diversitas International of the Western Pacific Asia (DIWPA) is a program to study the biodiversity in the western Pacific area. This area is unique on

---

\* This letter was used for soliciting participation in IBOY 2001. It is provided here only as a sample of the type of information required in such an invitation.

the earth in that it has forests distributed continuously from the northern to the southern boreal zone. A project focused on the forest ecosystems started in 1994, and will be carried out in IBOY 2001.

In addition to forests, DIWPA has recognized freshwater lakes and marine coasts as additional major ecosystems to study in this region. In the western Pacific area it is possible to study biodiversity of marine organisms from the coasts of Russia to New Zealand, again without any discontinuity along a latitudinal transect which encompasses temperate, subtropical and tropical areas.

One of the unique features of DIWPA is the opportunity to observe the biodiversity of sites along the latitudinal transect using the same sampling protocols. The methodology for the study of coastal biodiversity was discussed in DIWPA meetings held in Ohtsu city in November 2000 and in Shirahama in June 2001. A sampling framework was agreed upon and the final draft is expected to be published in December 2001. Only through this standardized strategy will it be possible to compare biodiversity from site to site, and determine the latitudinal pattern of biodiversity in the coastal zone on a global scale.

The detail of the methodology to be applied in the study of the DIWPA coastal zone is described in a separate paper. Briefly, the following points have been fixed.

1. Candidate sites for the field work were identified. At core sites, long-term monitoring of biodiversity exceeding 50 years is expected to be carried out. At satellite sites, field work will be done at least in the IBOY year, and additional work is optional. Areas where additional sites are needed were recognized. Final selection of the sites will be based on the response to a questionnaire to be distributed.

2. Two types of habitat, macroalgae-hard and seagrass-soft-substrate, of moderate size (shoreline extent of 20–200 m) have been chosen as the core study habitats. Sampling of other habitats such as mangrove, coral reef and unvegetated substrates are also recommended, but are optional. All macrophytes and conspicuous macrofauna (> 2 cm) such as holothurians, echinoids, decapod crustaceans and mollusks, were selected as the key organisms to be studied and identified in situ. Fishes are optional taxa but are recommended to be studied.

3. For the future study aimed at considering the smaller benthic organisms, samples for macrofauna and meiofauna will also be collected.

4. For the study of key organisms, a stratified random sampling strategy is to be adopted. Replicate quadrats will be sampled within each habitat

at high-, middle- and low-water positions in the intertidal zone and at depths 1 m, 5 m and 10 m subtidally. A detailed methodological protocol is attached.
5  The first year of fieldwork should be completed by the end of 2002.
6  A workshop will be held in late 2002 or early 2003 to discuss and analyse the results of IBOY sampling.
7  Data management will be implemented through OBIS (Ocean Biodiversity Information System), an international project .
8  Collaboration with a similar European project (MARS-BIOMARE) will be sought.

The emphasis of the DIWPA coastal zone project on measuring taxonomic diversity on such a large scale using the same protocol fits well with the goals of a recently developed international activity named CoML (Census of Marine Life: http://www.coml.org). The major goal of CoML is to assess and explain the diversity, distribution and abundance of marine species worldwide. This ambitious project was launched in 2000 with the support of several funding agencies, e.g. Sloan Foundation, New York. OBIS is the first project of CoML. Thanks to their keen attention, CoML decided to support the DIWPA coastal study as one of their 10 pilot projects.

The Japan Society for the Promotion of Science (JSPS) also recognized DIWPA as an activity worthy of support and in their new multinational project of marine science, IBOY DIWPA will be partially supported.

The fiscal resources of DIWPA as a whole are very limited – the activity should be supported mostly by grants obtained by each individual participant. However, because CoML recognizes DIWPA as a pilot project, it may be much easier for each participant to raise the required funds. In addition, those participants who have not been able to raise funds may seek direct support from CoML.

With the participation of more marine stations in the project, the regional biodiversity data obtained through DIWPA will be more extensive and precise. Your marine station is situated in a key locality for the project. Thus, I sincerely invite your participation in the DIWPA coastal module activity. If you agree to participate, please reply to the questionnaire attached.

Thank you for your kind consideration of the above.

# DIWPA IBOY CORE or SATELLITE SITE QUESTIONNAIRE FORM

## Contact details
Proposer Name:
Institution Name:
Postal address, tel, fax and e-mail:

## Proposed site
Site name:
Location (latitude, longitude):
Brief description of the site:

## Criteria for site evaluation

1. Is your site pristine with respect to anthropogenic stresses (relative to the conditions dominant in the region)?

    please mark X one of the following:

    ( ) Low
    ( ) Moderate
    ( ) High (= pristine)

    Justification:

2. Does homogeneous macroalgae-hard substratum habitat (shoreline extent 20–200 m) occur at your site?

    please mark X     ( ) yes   ( ) no

3. Does homogeneous seagrass-soft substratum habitat (shoreline extent 20–200 m) occur at your site?

    please mark X     ( ) yes   ( ) no

4. What other habitats are present at your site?

    please mark X any of the following which exist (and/or list others):

    ( ) mangrove
    ( ) coral reef
    ( ) unvegetated substratum – mud, muddy-sand, sand, gravel
    ( ) other:

5. What sources of environment and biodiversity information are already available for your site?

    please mark X any of the following which exist:

    ( ) unpublished data

( ) grey literature (e.g. reports)
( ) Primary literature

6  Is your study site protected by legislation that offers a guarantee that it will remain relatively pristine in the future?
   please mark X appropriate level of legislation:
   ( ) local
   ( ) regional
   ( ) national
   ( ) international
   please name category of protection (e.g. marine reserve):

7  Is your study site proximal to, or easily supported by your institution?
   please mark X    ( ) yes   ( ) no

8  What resources and facilities are available at your institution to support the proposed study?
   please mark X appropriately from the following:
   ( ) field sampling equipment
   ( ) diving equipment and expertise
   ( ) researchers with the ability to identify in situ:
       ( ) macroalgae
       ( ) seagrass
       ( ) large macrofauna
       ( ) Cnidaria
       ( ) Mollusca
       ( ) Decapoda
       ( ) Echinodermata
       ( ) Pisces
   ( ) sample storage
   ( ) library

# Appendix 3: List of acronyms

| | |
|---|---|
| BIOMARE | Implementation and Networking of large-scale long-term Marine Biodiversity Research in Europe |
| CoML | Census of Marine Life |
| MARS | EU Network of Marine Station |
| NAML | US National Association of Marine Laboratories |
| OBIS | Ocean Biodiversity Information System |

# Drafting Team

Yoshihisa Shirayama, Seto Marine Biological Laboratory, Kyoto University
Ashley A. Rowden, Marine Biodiversity Group, National Institute of Water & Atmospheric Research (NIWA)
Hiroshi Kawai, Research Center for Inland Seas, Kobe University
Hiroshi Mukai, Northern Biosphere Field Science Center, Hokkaido University
Allan J.K. Millar, The Royal Botanical Garden, Sydney
Jeong Ha Kim, Department of Biological Science, Sungkyunkwan University
Yoon Sik Oh, Department of Biology, Gyungsang National University
Zaidi Che Cob, School of Environmental and Natural Resource Science, National University of Malaysia
Asmida Ismail, School of Environmental & Natural Resource Sciences, Universiti Kebangsaan Malaysia
Susetiono, Research Centre for Oceanology, Indonesian Institute of Sciences (LIPI)
Husni Askab, Research Centre for Oceanology, Indonesian Institute of Sciences (LIPI)
Hong Zhou, Swire Institute of Marine Science, Department of Ecology & Biodiversity, University of Hong Kong
Xiaorong Tang, College of Marine Life, Ocean University of Qingdao
Dennis P. Gordon, Marine Biodiversity Group, National Institute of Water & Atmospheric Research (NIWA)

# Chapter 5: Research Methods to Initiate PABITRA: The Island Ecosystem Branch of DIWPA

*Chapter Editor: Dieter Mueller-Dombois*

## 5.1 Introduction

PABITRA, the Pacific-Asia Biodiversity Transect network, is the tropical island branch of DIWPA, the international network of DIVERSITAS in the Western Pacific and Asia (Yumoto 1999). The design of DIWPA includes a north-south oriented "Green Belt" from Russia via tropical East Asia to Australia and New Zealand, and a roughly parallel running "Blue Belt." The research focus along the "Green Belt" is on biodiversity relations in a number of selected forest and lake ecosystems. The research focus along the "Blue Belt" refers to biodiversity studies in selected coastline ecosystems.

PABITRA, the east-west ranging "Tropical Island Belt", intersects with several DIWPA sites in the continental islands of the Western Pacific (New Guinea, Borneo, the Philippines, and Taiwan). PABITRA's research focus combines horizontal and vertical approaches to ecosystem studies. The 'horizontal approach' implies an initial concentration on comparative biodiversity research of the indigenous upland and inland forests of the high-island archipelagoes as ecological reserves and watershed covers. At the same time, the 'vertical approach' implies cross-ecosystem studies from the watershed covers down to the fringing reefs (Mueller-Dombois et al. 1999).

A manual for biodiversity assessment of tropical island ecosystems is currently in preparation. It is planned as a 15 chapter volume of methods for island forests, stream ecosystems, agricultural areas, and coastal zone habitats, which on most volcanic high islands form a closely interacting human support system. A first draft was discussed at the DIWPA workshop in Sydney, July 4, 1999. At that time, I was asked to write a PABITRA contribution, which was subsequently published as Chapter 5 of IBOY-DIWPA.

Following are excerpts of the PABITRA document submitted to Kyoto. The full version also served as PABITRA progress report for 1999, of which other parts, such as the "current and planned projects", "organization of PABITRA", and "site selection criteria" are now on the PABITRA web-site: www.botany.hawaii.edu/pabitra/. A few new developments have been added to the concluding section.

## 5.2 The DIWPA/PABITRA Relationship

Dr. Takakazu Yumoto, Secretary of DIWPA at the Center for Ecological Research, Kyoto University, presented a paper at the Second PABITRA workshop/symposium in Taipei, November 1998. In this paper, Yumoto (1999) characterized DIWPA as "a non-governmental, apolitical, regional organization focussed on the integration of efforts and resources towards the study of the unique ecosystems of the western Pacific and Asia including the terrestrial Green Belt from Siberia to New Zealand and the marine Blue Belt along its eastern border (Inoue 1996), as well as the Pacific East-West Island Belt connecting the tropical archipelagoes whose biota are largely derived from continental islands in the western Pacific" (Mueller-Dombois 1998).

The mission of DIWPA is the promotion of sustainable management and utilization of biodiversity in the Pacific-Asia region. This mission statement is defined by 10 objectives:

1 Promote regional research of biodiversity
2 Develop joint projects at the regional level
3 Establish regional biodiversity networks
4 Encourage interchange of information among scientists
5 Conduct training courses pertinent for biodiversity
6 Provide scientific bases for developing common regional policies for biodiversity management and conservation
7 Establish a network of data bases
8 Develop capacity in biodiversity assessment and analysis
9 Organize meetings, workshops, and symposia periodically on current regional issues and concerns regarding biodiversity
10 Contribute to accomplishing the 10 Core Program Elements of DIVERSITAS' Operational Plan (Younes 1996)

These 10 core elements apply equally to PABITRA (Mueller-Dombois et al. 1999).

Yumoto outlined the more immediate goals for IBOY 2001 as:
1 Establishing an international network for the identification and preservation of biological specimens
2 Elucidating the effects of biodiversity on ecosystem functioning by inventorying and monitoring
3 Determining the implications of IBOY on research sites and standardized methodology

He added that IBOY 2001 would include a campaign to elevate public awareness of the critical importance of biodiversity conservation and research. This also should serve as a reminder of government decisions made at the 1992 Rio Earth Summit and Article 7 of the United Nations Convention on Biodiversity. This article requires signatory parties to "identify components of biodiversity important to conservation and sustainable use – and monitor, through sampling and other techniques, the components of biodiversity identified."

PABITRA is totally in line with these global concerns on biodiversity, the DIWPA mission statement, its 10 objectives, and the three or four more immediate aims of IBOY 2001.
1 In terms of establishing an international network for the identification and preservation of biological specimens, PABITRA will participate via the PSA Task Force on Biodiversity chaired by Dr. Allen Allison, Vice President for Research, B. P. Bishop Museum, 1525 Bernice Street, Honolulu, Hawai'i 96817, U.S.A. Tel. +(808) 848-4145; Fax +(808) 847-8252; e-mail: *allison@bishop.bishop.hawaii.org*. Dr. Allison will be in a position to advise and contact specialists for identification and preservation of almost all organism groups and taxa of relevance to Pacific biodiversity conservation.
2 Elucidating the effects of biodiversity functions in an island ecosystem context is one of the long-term goals of all PABITRA participants. For the short term of two years, from the Sydney workshop, July 1999 to IBOY 2001, the PABITRA Core Participants (listed under Organization of PABITRA) are considered the immediate resource persons contributing to IBOY 2001.
3 In terms of PABITRA sites, we pre-selected 20 proposed transect sites and 17 peripheral sites in Taipei (November 1998) and further scrutinized them in a brief PABITRA meeting in Honolulu (January 1999). These sites are listed in Mueller-Dombois et al. (1999).

## 5.3 Underlying Theories for PABITRA

The island biogeography theory of MacArthur and Wilson (1963, 1967) provides an underlying scientific inspiration for PABITRA (Fig.5.1). It predicts that a large archipelago far removed from a biotic source area arrives at a lower species equilibrium than a similarly large archipelago closer to a biotic source area. While the concept of species equilibria is highly debatable, it is generally accepted that there is a gradient of indigenous colonizer plants from enriched to impoverished as one travels from the species rich continental islands in tropical Asia into the oceanic archipelagoes of Polynesia. The Marquesas in Eastern Polynesia and the Hawaiian Islands in Northern Polynesia are recognized as the outlier landmasses of the Paleotropics (Barthlott et al. 1996). This west to east gradient of impoverishment of colonizer plants is well documented for mangrove species by Woodroffe (1987). It also becomes apparent from the canopy tree guilds comprising the inland and upland forests of the volcanic high islands. Inclusion of the ecological concepts of biome and succession will improve the theory of island biogeography (Mueller-Dombois 2001).

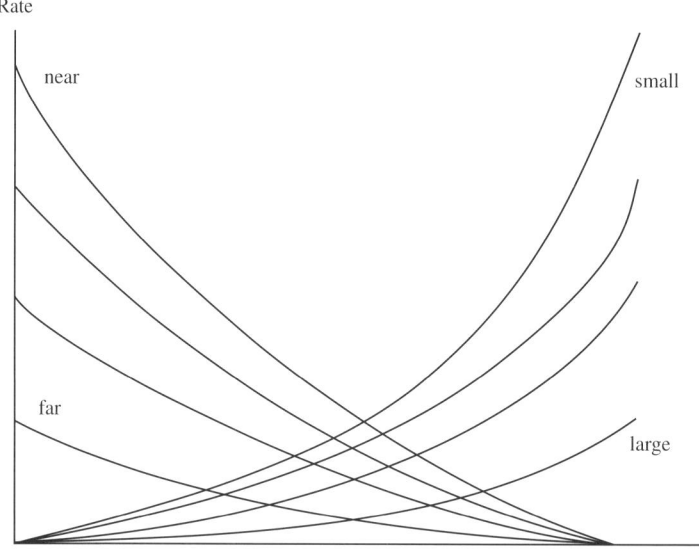

Figure 5.1: *The island biogeography model of MacArthur and Wilson (1967). It predicts, for example, that a large archipelago near a biotic source area arrives at a high equilibrium, whereas a similarly large archipelago far removed, arrives at a lower (impoverished) equilibrium.*

A recent Springer book synthesizing the vegetation of the tropical Pacific Islands by Mueller-Dombois and Fosberg (1998) can be used as a baseline document. This book treats the Pacific Island vegetation in a landscape perspective as ecosystem builders in the sense of the well-known British vegetation ecologist Tansley (1935). He stated that one can not understand the plant community without its environmental context. Tansley declared the community together with its habitat as an "ecological system." For this combined unit, he coined the abbreviated term "ecosystem." Tansley's term, launched in 1935, is now widely used and understood even by the non-specialist.

Mueller-Dombois and Fosberg (1998) consider the term "landscape" to be the geographic equivalent of the term "ecosystem." In this sense, they developed a formula for defining and describing vegetation as a function of six factors:

Vegetation = $f$(g, cl, d, fl, ac, e), where

$f$ = function,

g = geoposition (referring to geography, geology, geomorphology, and ground=soil),

cl = climate

d = disturbance regimes

fl = flora of the island area

ac = access potential of plants to become part of an ecosystem

e = ecological characteristics of plants and their role in the ecosystem

Overriding these six factors are the scales of space and time. Scaling is an essential element in understanding vegetation. But scaling does not affect the above formula, which can be applied at any scale from broad to detailed.

If one substitutes the factor fl (=flora) with b (=biota) in the formula, the whole community would need to be synthesized. This is a task for the future that, however, can hardly be done in one book on a Pacific-wide scale. Instead, it would require about 10 books, one for each island region.

Based on this formula, the vegetation of the Pacific Islands was described in 10 regions as outlined by the section index maps in Fig. 5.2. Their sequential numbers follow the general trend of biogeographic regionalization of the Pacific Island area as reviewed by Stoddart (1992).

The PABITRA concept evolved in part from the last chapter in the Mueller-Dombois and Fosberg book entitled "The Future of Island Vegetation." In the conclusions to that chapter, reference is made to DIWPA as a new research initiative, which includes Polynesia. Subsequently, the PABITRA initiative was established as explained above. PABITRA now is visualized as an east-west island transect network (Fig. 5.3) that complements the north-south transects visualized as the "Green Belt" and the "Blue Belt" of DIWPA.

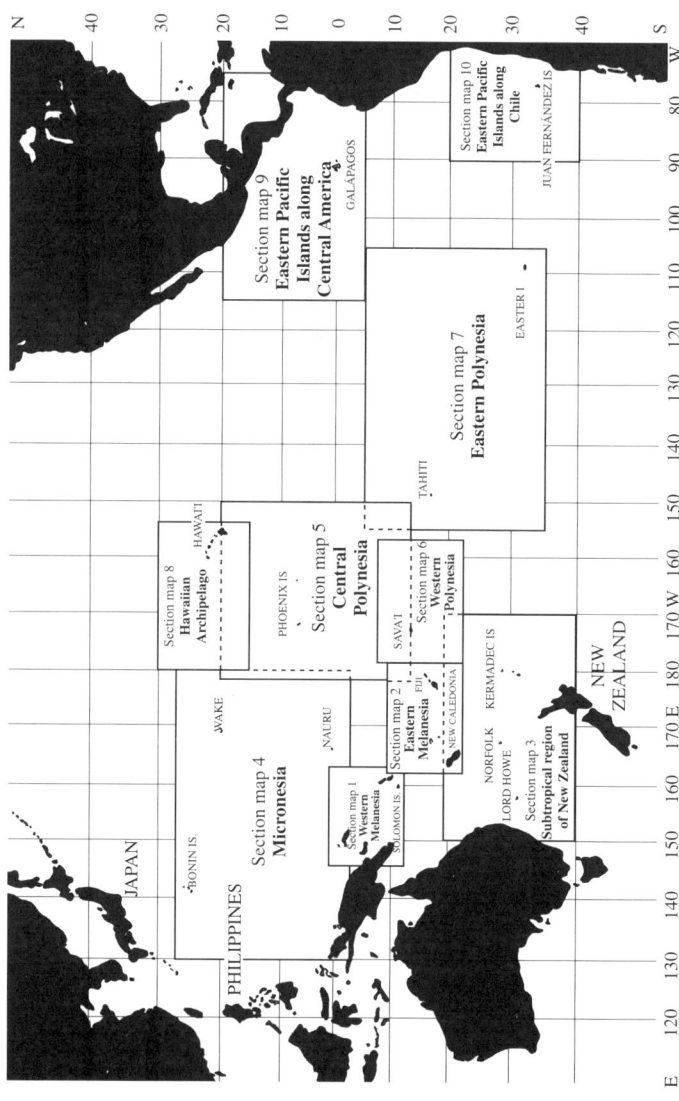

Figure 5.2: Pacific area map outlining ten island regions (1-10) in a forest biogeographical order. The regional maps are enlarged for each chapter in the vegetation book of Mueller-Dombois and Fosberg (1998).

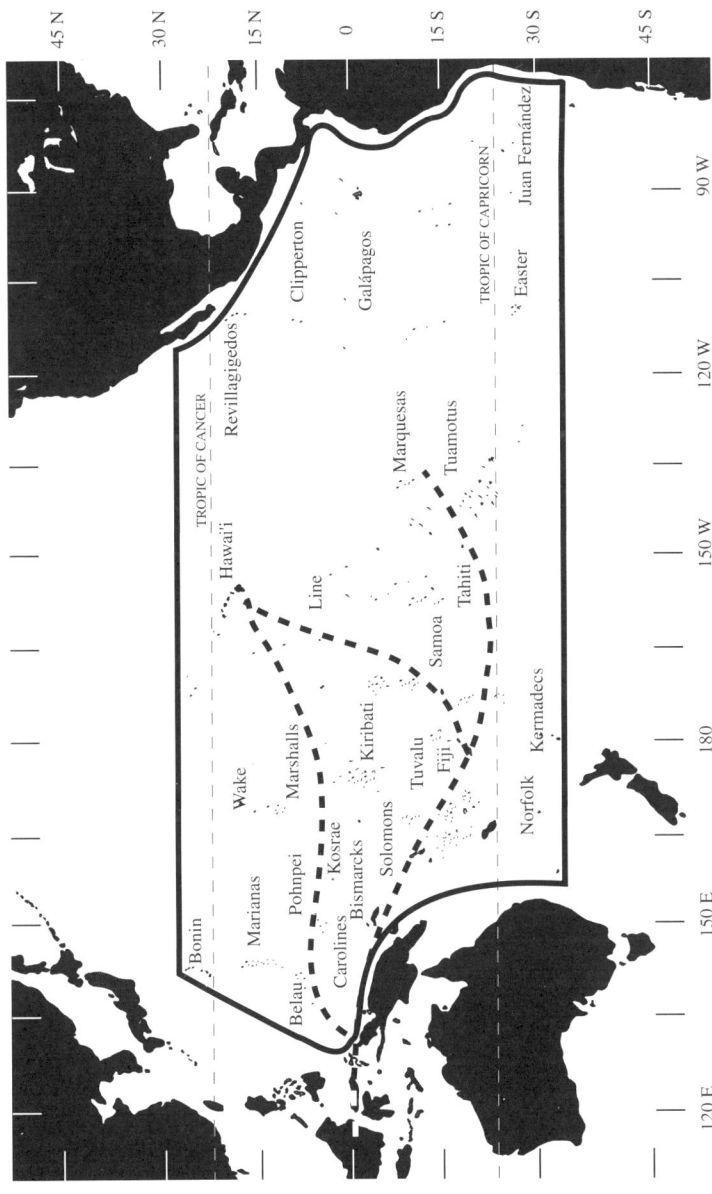

*Figure 5.3: The PABITRA (Pacific-Asia Biodiversity Transect) Network as currently envisioned. PABITRA represents the East-West Island Belt across the tropical Pacific that intersects with DIWPA's Green and Blue Belts in the Western Pacific.*

## 5.4 Geoecology and Environmental Gradients

The term geoecology derives from Geographical Ecology, a book written by MacArthur (1972). He emphasized research into the patterns of diversity, species and community distributions over wider areas as well as more local areas, such as the distribution of species and communities on mountain slopes. MacArthur wrote, "…to do science is to search for repeated patterns, not simply accumulate facts, and to do the science of geographical ecology is to search for patterns of plants and animal life that can be put on a map..."

Subsequently, Pielou (1979) in her book on biogeography defined geoecology *sensu* MacArthur as the "study of the recurrence of similar communities in similar habitats occupied by different sets of species." In other words, the principal concern of geoecology is the study of similar communities, members of the same biome, which grow in similar environmental settings but in biogeographically different areas. Geoecology in this sense is exactly the first approach of PABITRA, namely to focus on a comparative study of the indigenous upland/inland forests of the Pacific high islands as ocean-fragmented members of the same biome and to study them structurally and functionally as communities harboring most of the indigenous and endemic biodiversity and then also for their ecosystem services as watershed covers.

This approach differs from the study of environmental gradients. Environmental gradient studies focus on the changes in biodiversity as related to changing environmental control factors. These may be temperature, rainfall, seasonality, substrate age, disturbance regime, or other factors. Here, the sampling attempt is to place the study plots along the spatially changing control factor by holding all other factors as constant as possible. For example, for studying the effect of spatially changing temperature on biodiversity, which is the major environmental control factor, correlated with latitude, along DIWPA's "Green Belt", the forest plots should ideally be selected in such a way that soil age, elevation, water balance, and disturbance regime are kept closely comparable or equivalent.

## 5.5 Comparative Sampling of Upland Forests

DIWPA's Protocol 'A' requests that a one-hectare plot be established for each DIWPA forest site. This plot size is now widely used in species-rich tropical forests. In addition two ¼ ha satellite plots are suggested to be established in the same forest entity at about 5–10 km distance from the central plot. These satellite plots are added to assess some measure of homogeneity or within-forest variation of the selected entity. The procedure for establishing a one-hectare forest plot is relatively simple and straight forward, but labor intensive, particularly in uneven terrain. The procedure is described in detail in the Forest Ecosystem chapter (2) of this manual.

A scale diagram of the one-hectare DIWPA plot is shown in Figure 5.4 (upper right) in relation to different plot sizes used in other proven field-ecological studies. What these different plot sizes all have in common is the underlying assumption that they are representative of a larger forest entity. There are three practical methods for testing this assumption.

The first is to use aerial photographs to identify relatively homogenous forest entities or forest segments by preparing a preliminary vegetation map. This should be done with some ground truthing at rather large, i.e. detailed, scales such as 1:10 000–1:15 000.

The second method is to establish species/area curves. This is usually done by counting all species initially in a small plot and by enumerating additional species with each enlargement of the sampling area. Figure 5.5 displays some species/area curves established by Ashton (1965) for species-rich tropical rain forests in Borneo. As can be seen from these curves, the one-hectare DIWPA plot is a fair compromise for the lowest curve. A sample of 2 hectares would double the species count for the upper curve and still not quite satisfy the "minimal area requirement." This requirement stipulates that at least 80–90 % of all plant species in the forest map unit should be inventoried in the sample plot (Mueller-Dombois and Ellenberg 1974).

A third method is to establish satellite plots and to use these for similarity (or difference) comparisons by multivariate analysis techniques. To do this properly with the DIWPA plot layout, requires six ¼ ha plots, the four clustered together in the one-hectare plot, and the two satellite ¼ ha plots.

These three methods provide a common basis for all of the different plot sizes shown in Figure 5.4. The small relevé example represents a plot size often used in European vegetation studies. It is based on the "minimal area requirement." Daubenmire's (1968) standard plot size of 375 $m^2$ is similarly small. It proved successful in the mountainous topography of the Pacific Northwest in eastern Washington State. The larger plot sizes (0.6 ha, 2.8 ha, and 6 ha on Fig. 5.4) were used in more level or moderately undulating topography, where the vegetation was relatively homogenous over wider areas. In these larger plot studies, the primary concern was with the quantitative assessment of trees. To obtain a representative count of tree species and their populations, the number of individuals counted is a more critical parameter than the area covered. An

## Biodiversity Research Methods

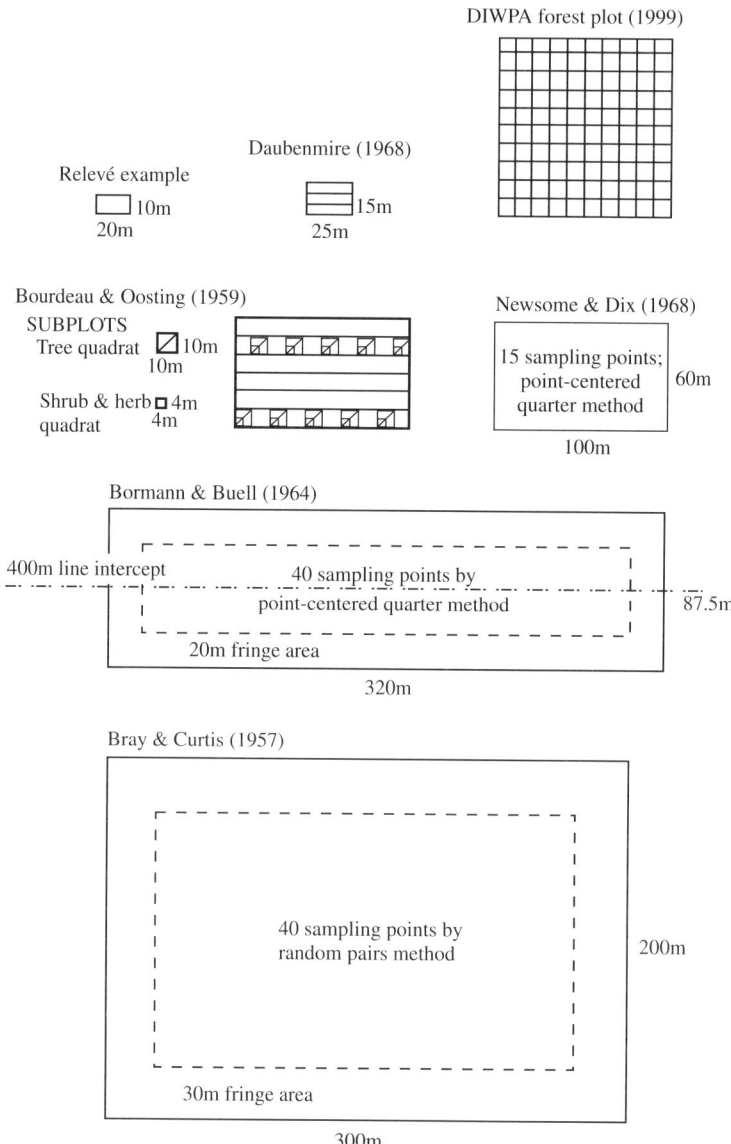

*Figure 5.4: Comparative plot sizes used for vegetation studies in different landscapes. At upper right, a one-hectare standard plot with 100 contiguous count-plots as suggested for DIWPA forest ecosystem studies.*

### Research Methods to Initiate PABITRA

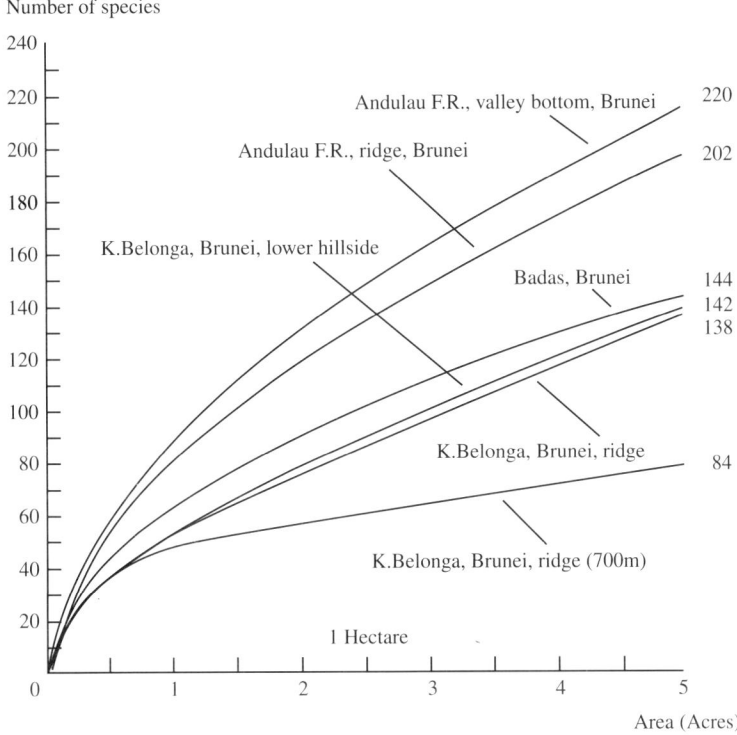

*Figure 5.5: Species/area curves for sites in Brunei. (Redrafted with slight modifications from Ashton 1965).*

adequate count/tree species usually requires enumeration of at least 30–50 individuals. In Bourdeau and Oosting's (1959) study, this was done in $10 \times 10$ m sub-quadrats, similar to those suggested for the DIWPA forest plots. In the other three studies, two types of distance methods, the point-centered quarter and random pair's methods were used to quantify tree species. The latter methods are usually faster and more elegant than the count-plot method. If applied properly, all of these methods and plot sizes produce comparable results. Proper application implies understanding the methods, the type of vegetation and the objectives of the study. If the vegetation is patchy at large (detailed) scales, small plot sizes are suggested. If the vegetation is homogenous over broad areas at similarly detailed scales, larger plot sizes are preferable.

Objectives: To survey the plant biodiversity of an area, a certain area coverage is required. DIWPA's minimum area coverage is one-and-a-half hectares. This will be accomplished in one hectare and two ¼ ha plots. The same area coverage (15 000 m$^2$) can be achieved with 75 relevés, each of 200 m$^2$, often considered a reasonable sample. In a rain forest study on the island of Hawai'i (Mueller-Dombois et al. 1980), we sampled 17 200 m$^2$ with 43 relevés, each of 400 m$^2$. This 1.72 ha combined plot sample gave a reasonable ecological representation of forest vegetation extending over 80 000 ha.

Monitoring, defined as repeated inventories in DIWPA's Protocol 'B', requires at least one fixed point in a target area. This implies the establishment of permanent sample plots so that repeat-inventories can be done at certain time intervals in the same locations. This is not easily accomplished. It requires well-chosen landmarks for proper relocation and frequent maintenance of the plot marker system. The recently developed global geographic positioning system (GPS) is a great technological breakthrough for this task. But individual trees should also be numbered and labeled for re-measurement, and tree labels must be renewed periodically.

Permanent plots can be carefully chosen as a subset of the initial inventory plots. In the Hawaiian rain forest study 25 of 43 plots were established as permanent plots. They were successfully relocated and re-measured at approximately 5-yr intervals (Jacobi et al. 1983, 1988). This monitoring study is continuing under the PABITRA program.

## 5.6 Sampling of Upland forests to Coastal Habitats

In addition to the horizontal transect design of PABITRA as depicted on Figure 5.3, PABITRA will include a number of vertical transects to connect an island's upland/inland forests to its lowland ecosystems and coastal zones. For this ecosystem cross-cutting approach, we will use surface freshwater flow, precipitation and temperature gradients as unifying parameters. In this upland/lowland design, we will study the inter-connectedness of island ecosystems. Freshwater is one of the most important renewable resources on each island. Most lowland ecosystems, the agricultural lands, tree gardens, freshwater marshes, mangroves, estuaries, coastal strips, lagoons, sea-grass beds, and fringing reefs are affected in various ways by the freshwater flow. Our main hypothesis is that the appropriate functioning of a native, self-reproducing forest watershed cover will be of significance to the health and appropriate functioning of all lowland ecosystems. This relates to appropriate nutrient cycling, pollution and erosion control. In fact, the interconnectedness of the lowland and coastal ecosystems to the upland and inland forests, has

Research Methods to Initiate PABITRA

traditionally been utilized by many Pacific Island societies as their human support system. The Hawaiians, for example, use the term "ahupua`a" for this system and recognize three vertically arranged zones, the "wao akua" (the inland or upland forest zone, the realm of the gods), the "wao kanaka" (the zone where humans are involved in agricultural activities), and the "kahakai", the coastal zone with its beaches, estuaries, lagoons, salt marshes, fringing reefs, and fish ponds. Hawai`i had no mangroves, only salt marshes, which later were quickly invaded after the introduction and planting of a few mangrove trees. The indigenous ahupua`a management system, which served successfully for many generations as the human support system, has never been investigated scientifically. PABITRA will attempt to do this by means of an integrated multi-disciplinary approach together with scientifically trained indigenous islanders and land managers. This upland/lowland segment of PABITRA also addresses the cross-cutting research elements of the DIVERSITAS program as outlined by Younes (1996). PABITRA will particularly be concerned with the human dimension of biodiversity, i.e. cross-cutting research element number 10 of the DIVERSITAS program.

An important design aspect is that vertical transects will be laid out wherever possible in both totally human modified and less modified, near-natural landscapes. This dual aspect is depicted on the diagram (Fig. 5.6) of an idealized high island. Here also the change in temperature, from the cool uplands to the warm and hot lowlands, serves as an important biodiversity control factor. Moreover, seasonality and disturbance regimes can be built into the sampling design, similar to the DIWPA "Green Belt" approach.

A methodology for vertical transect studies has been outlined in the IUBS Manual of Methods for Mountain Transect Studies, which was prepared for the IUBS sponsored "Decade of the Tropics" Program by van der Hammen, Mueller-Dombois, and Little (1989). This manual will be used as a supplement to the PABITRA protocol.

## 5.7 Survey of Selected Taxa: The Hawai'i IBP Example

During the 1970s, a multi-disciplinary biological inventory was done in Hawai'i as a contribution to the International Biological Program (IBP). The IBP was the first IUBS promoted multi-disciplinary biology program with wide-reaching international participation. DIVERSITAS can be considered its successor.

After developing some inspiring research hypotheses, based on prior field research, reviewed in a baseline document for Hawai'i Volcanoes National Park

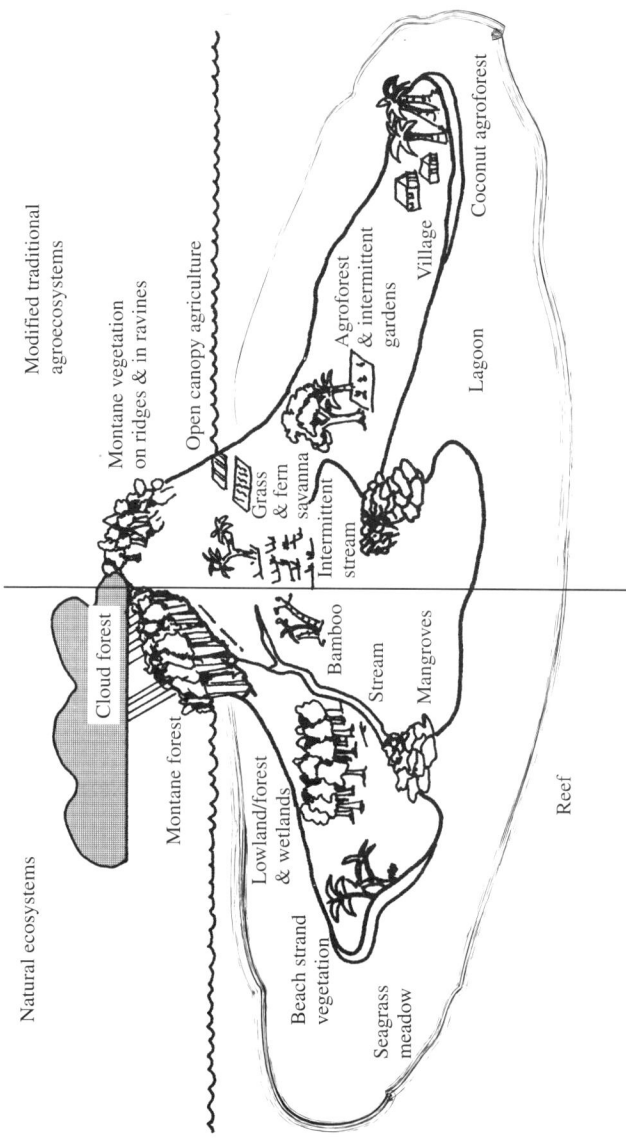

*Figure 5.6: Idealized volcanic high island with near-natural landscape on left side and strongly human-modified landscape on right side. PABITRA will study these landscapes by vertical transects, connecting mountain tops and upland/inland forests to the coastal ecosystems. (Source OTA, US Govt. 1987).*

(Doty and Mueller-Dombois 1970), the Hawai'i IBP proposal focussed on biodiversity research in an ecosystem context. We decided as a multi-disciplinary team to do a mountain transect and a study of a species-rich, well preserved, upper montane rain forest. The Hawai'i IBP can be considered as satisfying DIWPA Protocol 'C'. The inventory of biodiversity was combined with an analysis of micro-environmental heterogeneity at many locations, and the survey was repeated a few times over a short (5-year) time frame.

The sample design for the mountain transect involved 14 focal sites, from closed rain forest at 1220 m altitude to the upper limit of sparse alpine scrub at 3050 m elevation on Mauna Loa (Fig. 5.7). For the study of the upper montane rain forest, we established an 80 ha plot at 1600 m elevation in the Kilauea rain forest.

Fourteen organism groups were studied along the transect, namely plants, birds, rodents, ectoparasites on rodents, canopy arthropods, arthropods on reproductive organs of trees, wood-boring Cerambycid beetles, selected detritophages (Diptera: Sciaridae = fungus-gnats), bait-trapped Drosophilids, litter-inhabiting Diptera, soil arthropods, terrestrial algae, soil and leaf fungi.

This assemblage of organism groups (belonging to many taxa) was selected because a number of biologists (in this case 22 specialists) familiar with the above organism groups, were willing to work together in a multi-disciplinary team. Important groups, such as spiders and snails, were not studied because researchers for these groups did not become available during this time-limited program. Similarly, it would have been of value to study the bryophytes and lichens among botanical organisms, but time and expertise limitations prevented the inclusion of these groups.

Working together as a team meant that each team member agreed to inventory his or her organism group at the focal sites as to species present together with an assessment of their population quantities. The analysis became an ecological study only when the identified species were also quantified by a measure suitable for that organism group. In many cases, for example for soil algae, canopy arthropods and other transient or ephemeral groups, this meant repeat sampling at the focal sites. This was necessary also because of seasonal changes along the transect. For this, a phenological study component was carried out along the transect. This, together with climatic data and seasonal variations in some organism groups, provided for an analytical procedure of short-term temporal variation, separate from the spatial analysis of organism groups along the mountain transect.

The term "focal site" was used since each organism group has its own area-requirement for adequate sampling. For example, plants were sampled in three

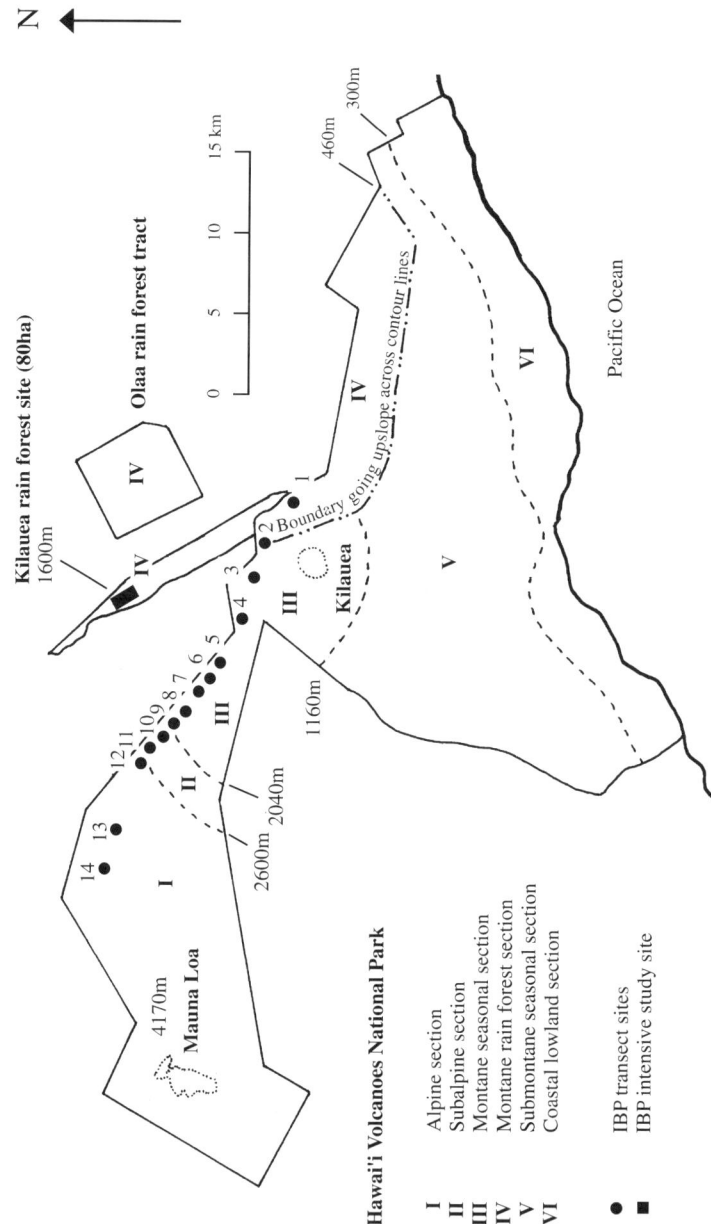

Figure 5.7: Map of Hawai'i Volcanoes National Park showing design of the Mauna Loa transect sites 1-14 and location of the 80 ha forest dynamics plot as sampled by multi-disciplinary teams during the Hawai'i IBP (From Mueller-Dombois et al. 1981).

or four relevés (each 400 m²) clustered at each focal site. Rodents were captured along 570 m trap lines, birds were censured over areas much larger than 1 ha/ focal site, soil arthropods were captured in a number of randomly distributed pitfall traps at the focal sites, etc. These data were then used to answer the hypotheses on species and community distribution that were stipulated for this mountain transect analysis. Details are published in US/IBP Synthesis Volume 15 (Mueller-Dombois, Bridges, and Carson 1981) and in 77 Technical Reports listed in the back of that book. All Hawai`i IBP Technical Reports are available on microfiche (or as hard copy on loan) from the Hamilton Library at the University of Hawai`i at Manoa in Honolulu, HI 96822.

## 5.8 Assessing Biodiversity Functions: Another Example from the Hawai'i IBP

The 80 ha Kilauea rain forest plot (whose location is shown in Fig. 5.7) was inventoried with a view to access the function of biodiversity in an ecosystem context. There are a good number of biodiversity functions, which can be derived from ecologically based inventories, but only two will be emphasized here. The first relates to an application of the guild concept, the second to an assessment of forest dynamics patterns.

### 5.8.1 Sampling protocol of a large (80 ha) forest plot

The 80 ha plot located in a (800 ha) homogenous tract of montane rain forest, was considered large enough to assess both of these biodiversity functions. The plot design is shown in Figure 5.8. The 80 ha plot was inside a forest reserve, protected by a fence from invasion by cattle, and located below an abandoned (never used) logging road. The plot was clearly tied to these landmarks. It can be relocated at any time if so desired. We established a climatic station at the NW plot corner, which was operated for the duration of the 5-year Hawai'i IBP project. As indicated in Figure 5.8, we sampled biodiversity along four 1000 m long transects, equally spaced, 200 m apart, along the 800 m base line. Five plot markers were established along each transect at 200 m intervals, resulting in 20 plot markers. These transect and reference points served for all sampling programs. Transect widths, plot sizes, and intervals were adjusted to each investigator's needs.

The basic sampling unit for the plant biodiversity survey was a belt transect or rectangular plot of $6 \times 100$ m (Fig. 5.9). Each of these 600 m² rectangular plots was subdivided for quantitative assessment. Undergrowth, including tree

## Biodiversity Research Methods

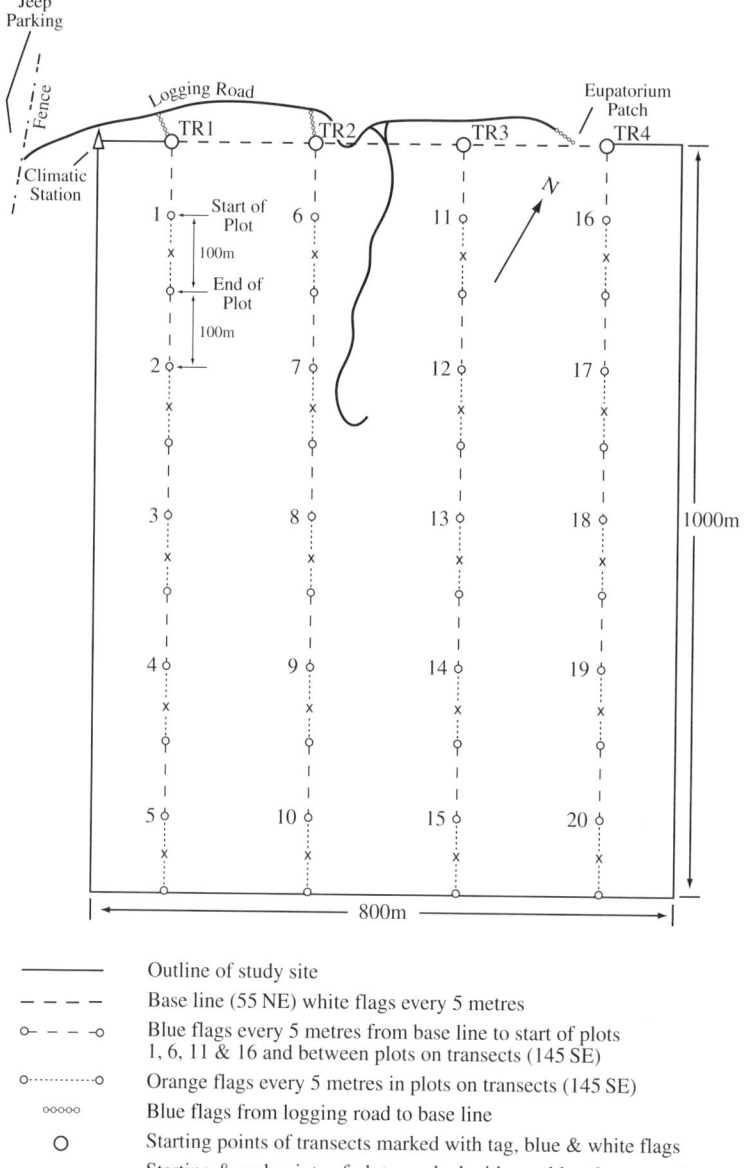

*Figure 5.8: Details of the 80 ha forest dynamics plot in the Kilauea rain forest. Four major transects (TR1, TR2, TR3, TR4) each with four plot starting points (1-20) were marked in the field at distances and directions shown (From Mueller-Dombois et al. 1981).*

regeneration, was assessed in 3 × 5 m subplots, and trees exceeding 2.6 cm dbh (at 1.45 m height) were counted by species and sizes in twenty 6 × 5 m subplots.

For compatibility with the DIWPA Forest Protocol, in the PABITRA forest studies we will use 10 m (instead of 6 m) wide belt-transects. Thus, the basic tree-count plots will be 10 × 10 m = 100 m$^2$ as recommended for the DIWPA forest plot (Fig. 5.4). This will allow for the direct comparison of tree species frequency counts, a measure that is dependent on quadrat size.

Forest animals were inventoried in the same sample units. Tree borers and bark beetles were studied repeatedly along all four transects by inspecting suggestive scars on trunks and fallen trees and by searching for free-moving specimens in flight and on foliage. Small mammals were sampled along two 1000 m transects monthly for two years, alternating among the four transects so that each was sampled bimonthly; snap and live traps for introduced rats were placed at 5 m intervals with every third one a snap trap. Birds were surveyed repeatedly by slowly walking along each transect with frequent stops using Emlen's method. The activity of feral pigs was monitored periodically by

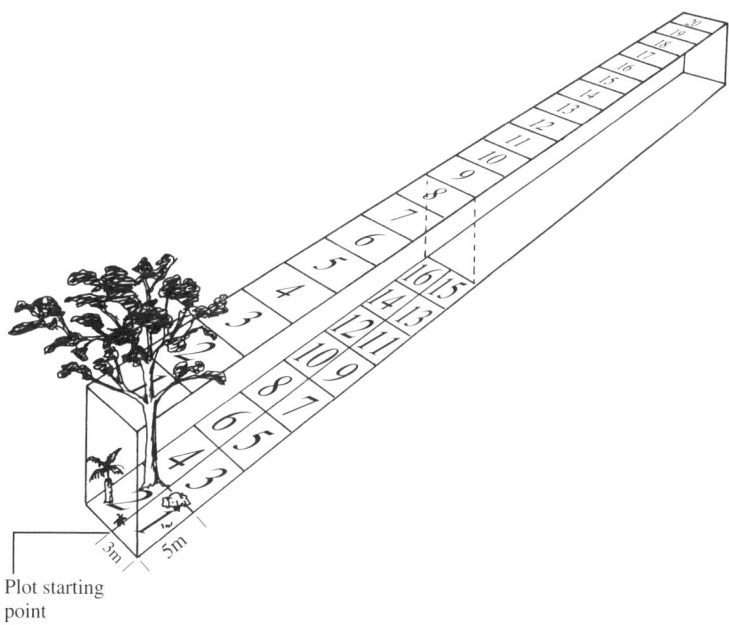

Plot starting point

*Figure 5.9: Basic sampling unit of plant-biodiversity survey, a long, 6 × 100 m belt-transect, with subplots adapted in size for quantifying tree and undergrowth vegetation (From Mueller-Dombois et al. 1981).*

recording activity signs (rooting, wallowing, freshly felled tree ferns, faeces, etc.) along each transect. Foliar arthropods were sampled in the central area of each transect at plots 3, 8, 13, and 18 on the dominant trees (*Acacia koa* and *Metrosideros polymorpha*) by bimonthly fogging of their crowns with synergized pyrethrum.

### 5.8.2 Application of the guild concept

All organism groups containing more than a few species were analyzed in guilds (or life form groups) and presented in guild spectra. As an example, the guild spectrum for the tree-associated arthropods is shown in Figure 5.10. The functional roles of these arthropods are differentiated first into primary consumers (herbivores) and secondary consumers (predators) on each of the dominant tree species. Their more detailed functional roles are recognized as 12, namely, defoliators = De, sapsuckers = Sa, seed predators = Se, twig borers = Tw, flower pollen and nectar feeders or pollinators = Fl, parasites = Pa, predators = Pr, fungivores = F, detrivores of plant and animal matter on host tree = D, detrivores in dead wood = W, leaf miners = L, and transients, which include perching arthropods and those with undetermined ecological roles. This guild spectrum is supported by a checklist of 177 arthropod taxa. Of these, 79 % were native taxa on *Metrosideros* and 69 % were native on *Acacia koa*. These were collected with different frequencies during repeated monitoring as shown by the height of the bars in Figure 5.10.

### 5.8.3 Assessment of forest dynamics patterns

During the quantitative vegetation survey, a few more or less distinct and repeating patterns became apparent in the otherwise homogeneous forest. These related primarily to canopy openings created through death and tree falls of the overstorey tree species, *Acacia koa*. We could recognize recently created gaps and older gaps. The older gaps were mostly filled with tree ferns (*Cibotium glaucum*) and other native trees, which rarely reach into the upper canopy (*Cheirodendron trigynum, Ilex anomala, Coprosma rhynchocerpa, Pelea = Melicope* spp.) and including occasional young *Acacia koa*. Another gap type was occupied mostly by saplings or young trees of *Metrosideros polymorpha*, which represents the reproduction of the second canopy species in this forest. These three gap types were encountered along the transects and in several locations between the transects. A few large-scale profile diagrams were prepared along plots 1, 5, 16, and 20, which further supported these patterns. A colour photo of the 80 ha plot taken vertically from a low-flying aircraft was obtained. On this large scale photo (1:1500) it was possible to identify

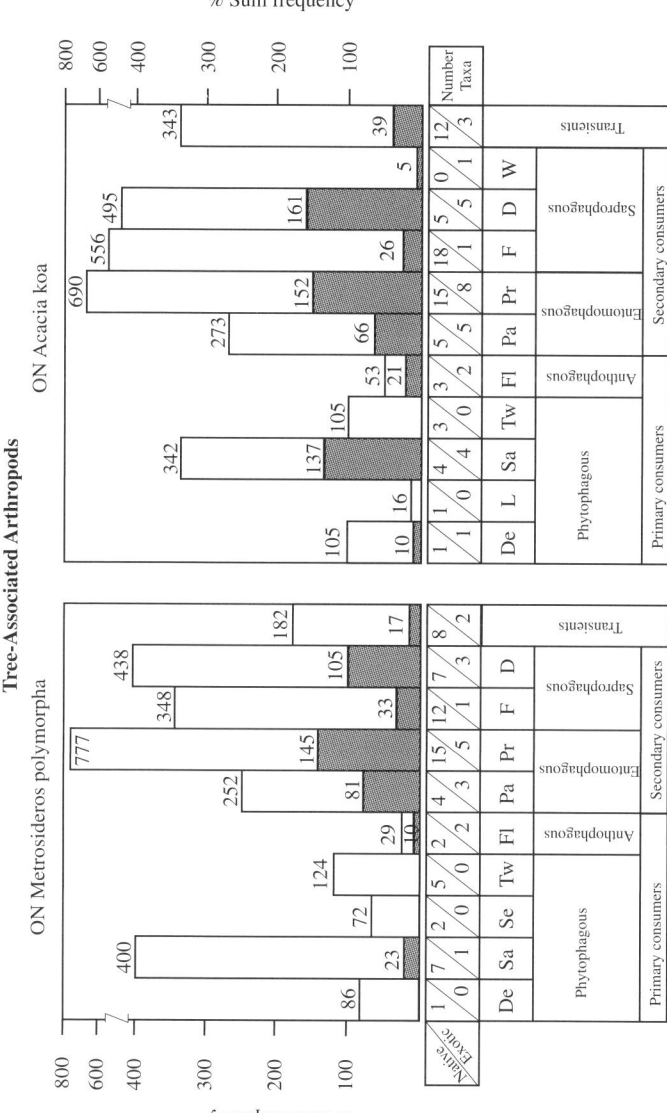

Figure 5.10: Guild spectrum of two arthropod communities in the Kilauea rain forest. For explanation of symbols see text (After Gagne and Howarth 1981).

individual crown outlines. The photo was used to map the four patterns recognized initially on the ground. The resulting plot map is here reproduced as Figure 5.11.

Analysis of the coverage of each pattern revealed that 75 % of the plot area was covered by the dominant tree crowns, only 1 % by recent windfalls of canopy trees, 17 % of the gap area was occupied mostly by tree ferns and 7 % gap area was dominated by *Metrosideros polymorpha* trees of mostly low-and intermediate stature.

Clearly, this large-scale map is a dynamic map. It shows different stages of forest development as related to the functioning of biodiversity in this ecosystem. The functioning is further related to the aging patterns (senescence reached in many individuals of *Acacia koa*) and the disturbance regime, i.e. 5 to 7 kona storms during the cooler season of each year. The two dominant canopy species in this forest are pioneers requiring canopy gaps for successful regeneration (Cooray and Mueller-Dombois 1981).

It is of some interest here to note that the latest newsletter of CTFS (the Center for Tropical Forest Science) includes an article on the advantages of incorporating plots smaller and larger than 1 ha into the sampling design of tropical forests (Losos 1999). The smaller plots, when spread out over a larger area, provide for measures of forest homogeneity versus heterogeneity, and they are less labor intensive. The larger plots, now suggested to cover 50 ha, are named by a new acronym FDP plots, i.e. forest dynamic plots. Researchers discovered that the standard one-hectare plots were too small to detect the more important patterns of forest dynamics. Thus, the Hawai'i IBP plot of 80 ha, described above, can be considered a proto-type FDP established in a Hawaiian rain forest already in 1972.

## 5.9 PABITRA Scope for IBOY 2001

In Sydney (during the 19th Pacific Science Congress, July 1999), suggested PABITRA transect sites were identified for the following island areas: Irian Jaya (Indonesia), Papua New Guinea (Bowutu Mts., Baitabag and Huon Peninsula), Taiwan (five LTER sites), Babeldaop (Republic of Belau), Yap Island (Federal States of Micronesia), Pohnpei (FSM), Kosrae (FSM), New Britain (Bismarck Archipelago), Solomon Islands (Kolombangara and Lauru), Fiji (Mt. Tomaniivi), Samoa (Savai'i, Western Samoa and Ta'u, American Samoa), Cook Islands (Rarotonga), Society Islands (Tahiti), Marquesas Islands (Nuku Hiva), Hawaiian Islands (Maui, particularly West Maui and East O'ahu). Among the

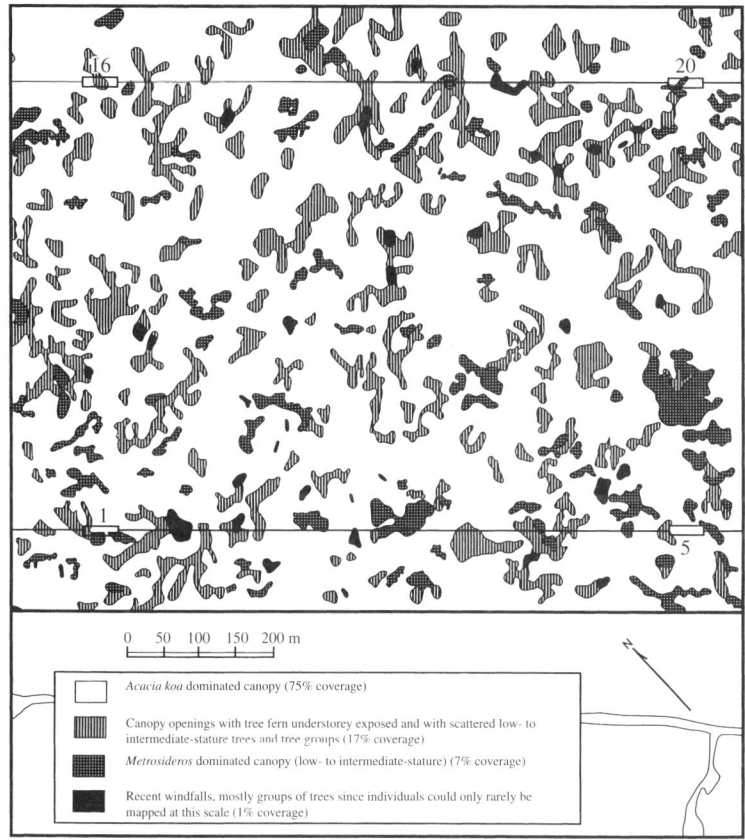

*Figure 5.11: Vegetation map of the 80 ha IBP study plot in the Kilauea rain forest, derived from a color air photo taken July 1974. Original map scale before reduction was 1:1500 (After Cooray and Mueller-Dombois 1981).*

peripheral sites, PABITRA contributions were made from: the Bonin Islands (Chichi Jima), the Galapagos Islands (a transect cutting across four islands, from young to old), the Juan Fernandez Islands (Robinson Crusoe Island=Masatiera: a proposed mountain transect and vegetation map). These 15 island areas along the horizontal transect system and the three peripheral ones will be further scrutinized for environmental comparability and their suitability as PABITRA sites.

A fourth PABITRA workshop/symposium was held in Nagano, July 2000, during the 43rd IAVS (International Association of Vegetation Science) Meeting. The theme was "Tasks for Vegetation Ecology in the PABITRA Net." Thirteen oral contributions were presented. Their abstracts are available on the PABITRA website *www.botany.hawaii.edu/pabitra/*.

The Council of the Pacific Science Association (PSA) has accepted PABITRA officially as the new ecosystem program in the PSA Task Force on Biodiversity. This also means that future PSA Congresses and Inter-Congresses will become the major meeting places for PABITRA symposia and workshops. The next PSA Inter-Congress is scheduled for June 1–6, 2001 in Guam. This coincides with IBOY 2001, and we plan sessions on "Science for Ecosystem Conservation" and "Methodology." The next full PSA Congress (the 20th) is planned to be in Bangkok, Thailand, March 17–21, 2003.

# References

Ashton, P.S. 1965 Some problems arising in the sampling of mixed rain forest communities for floristic studies. In: *Symposium on Ecological Research in Humid Tropics Vegetation* (1963). Sponsored by Gov. of Sarawak and UNESCO Science Cooper. Office for SE Asia, Tokyo Press, Itabashi, Tokyo, pp. 235–40.

Barthlott, W., Lauer, W., & Placke, A. 1996 Global distribution of species diversity in vascular plants: Towards a world map of phytodiversity. *Erdkunde* 50: 317–27.

Bormann, F.H. & Buell, M. F. 1964 Old-age stand of hemlock-northern hardwood forest in central Vermont. *Bull. of Torrey Bot. Club* 91(6): 451–65.

Bourdeau, P.F. & Oosting, H.J. 1959 The maritime live oak forest in North Carolina. *Ecology* 40(1): 148–52.

Bray, J.R. & Curtis, J.T. 1957 An ordination of the upland forest communities of southern Wisconsin. *Eco. Monogr.* 27: 325–49.

Cooray, R.G. & Mueller-Dombois, D. 1981 Profile diagrams and map; population structure of woody species. In: Mueller-Dombois, D., Bridges, K.W., & Carson, H.L. (eds.) *Island Ecosystems: Biological Organization of Selected Hawaiian Communities*, US/IBP Synthesis Series 15, Hutchinson-Ross, Woodshole, Massachusetts, pp.251–68.

Daehler, C.C. & Carino, D.A. 1999 Threats of invasive plants to the conservation of biodiversity. In: Chou, C.-H., Waller, G.A. & Reinhardt, C. (eds.) *Biodiversity and Allelopathy: From Organisms to Ecosystems in the Pacific.* Academia Sinica, Taipei, pp.21–7.

Daubenmire, R.F. 1968 *Plant Communities: A textbook of plant synecology*. Harper & Ross, New York.

Doty, M.S. & Mueller-Dombois, D. 1970 *Atlas for Bioecology Studies in Hawaii Volcanoes National Park*. College of Trop. Agric. Hawaii Agric. Exp. Sta. Miscell. Publication 89. (Republication of 1966 Hawaii Botan. Sci. Paper No. 2).

Doyle, M.F. 1999 Summary and conservation status of the native gymnosperms of the oceanic islands of the tropical SW Pacific (Fiji, Tonga, Solomon Islands, and Vanuatu). In: Chou, C.-H. Waller, G.R. & Reinhardt, C. (eds.) *Biodiversity and Allelopathy: From Organisms to Ecosystems in the Pacific*. Academia Sinica, Taipei, pp.215–20.

Gagné, W.C. & Howarth, F.G. 1981 Arthropods associated with foliar crowns of structural dominants. In: Mueller-Dombois, D. Bridges, K.W. and Carson, H.L. (eds.) *Island Ecosystems: Biological Organization of Selected Hawaiian Communities*. US/IBP Synthesis Series 15. Hutchinson Ross, Woodshole, Massachusetts, 275–90.

Inoue, T. 1996 Biodiversity in Western Pacific and Asia and an action plan for the first phase of DIWPA. In: I. M. Turner et al., (eds.) *Biodiversity and the Dynamics of Ecosystems*. DIWPA Series Vol. 1, pp.13–31.

Jacobi, J.D., Gerrish, G., & Mueller-Dombois, D. 1983 'Ohi'a dieback in Hawai'i: vegetation changes in permanent plots. *Pac. Sci.* 37: 327–37.

Jacobi, J. D., Gerrish, G., Mueller-Dombois, D., & Whiteaker, L. 1988 Stand-level dieback and regeneration in the montane rain forest of Hawai'i. *GeoJournal* 17: 193–200.

Kitayama, K. 1996 Patterns of species diversity on an oceanic versus a continental island mountain: a hypothesis on species diversification. *J. of Vegetation Science* 7: 879–88.

Kitayama, K. and Mueller-Dombois, D. 1997 Workshop on biodiversity transect held at the 8th Pacific Science Inter-Congress, Fiji. *Pac. Sci. Assoc. Inform. Bull.* 49: 10–11.

Losos, E.C. 1999 Small plots complimenting large. *Newsletter of the Center for Tropical Forest Science*. Smithsonian Trop. Res. Inst., Washington, D.C. Summer 1999 Issue.

MacArthur, R.H. & Wilson, E.O. 1963 An equilibrium theory of insular biogeography. *Evolution* 17: 373–87.

MacArthur, R.H. & Wilson, E.O. 1967 *The Theory of Island Biogeography*. Monographs in Population Biology 1. Princeton University Press, Princeton.

MacArthur, R.H. 1972 *Geographical ecology: patterns in the distribution of species*. Harper and Row, New York.

Mueller-Dombois, D. 1998a Vegetation and ecosystem research for biodiversity conservation in the Pacific islands. *Pac. Sci. Assoc. Inf. Bull.* 50: 1–10.

Mueller-Dombois, D. 1998b Plant biodiversity in tropical ecosystems across the Asia-Pacific region. In: Chou, C.-H. & Shao, K.T. (eds.) *Frontiers in Biology: The Challenges of Biodiversity, Biotechnology, and Sustainable Agriculture.* Proceed. of the IUBS Symposium. Academia Sinica, Taipei, pp.105–13.

Mueller-Dombois, D. 1999 Biodiversity and environmental gradients across the tropical Pacific Islands: A new strategy for research and conservation. *Naturwissenschaften* 86: 253–261.

Mueller-Dombois, D. 2001 Island biogeography. *Encyclopedia of Biodiversity*, Vol. 3: 565–80.

Mueller-Dombois, D. & Ellenberg, H. 1974 *Aims and Methods of Vegetation Ecology.* John Wiley & Sons, New York.

Mueller-Dombois, D., Bridges, K.W., & Carson, H.L. (eds.) 1987 *Island Ecosystems; Biological Organization in Selected Hawaiian Communities.* US/IBP Synthesis Series 15. Hutchinson Ross, Woodshole, Massachusetts.

Mueller-Dombois, D. & Kitayama, K. 1996. Research hypotheses for DIWPA cooperation. In: Turner, I.M. Diong, C.H. Lim, S.S.L. & Ng, P.K.L. (eds.) *Biodiversity and the Dynamics of Ecosystems.* DIWPA Series Vol. 1, pp.33–7.

Mueller-Dombois, D. & Fosberg, F.R. 1998 *Vegetation of the Tropical Pacific Islands.* Springer-Verlag, Heidelberg, N.Y.

Mueller-Dombois, D., Thaman, R.A., Juvik, J.O., & Kitayama, K. 1999 The Pacific-Asia Biodiversity Transect (PABITRA). A new Conservation Biology Initiative. In: Chou, C.-H., Waller, G.A. & Reinhardt, C. (eds.) *Biodiversity and Allelopathy: From Organisms to Ecosystems in the Pacific.* Academia Sinica, Taipei, pp.13–20.

Newsome, R.D. & Dix, R.L. 1968. The forests of the Cypress Hills, Alberta and Saskatchewan, Canada. *Am. Midland Naturalist* 80: 118–85.

Pielou, E.C. 1979 *Biogeography.* John Wiley & Sons, N.Y.

Tansley, A.G. 1935 The use and abuse of vegetational concepts and terms. *Ecology* 16: 284–307.

Stoddart, D.R. 1992 Biogeography of the tropical Pacific. *Pac. Sci.* 46: 276–93.

Van der Hammen, T., Mueller-Dombois, D. & Little, M.A. 1989 *Manual of Methods for Mountain Transect Studies. Comparative Studies of Tropical Mountain Ecosystems.* Intern. Union of Biol. Sci. (IUBS), Paris, France.

Woodroffe, C.D. 1987 Pacific island mangroves: distribution and environmental settings. *Pac. Sci.* 41: 166–85.

Younes, T. 1996 Biodiversity science: issues and challenges. In: Turner, I.M.,

Diong, C.H., Lim, S.S.L. & Ng P.K.L. (eds.) *Biodiversity and the Dynamics of Ecosystems*. DIWPA Series Vol. 1, Kyoto, pp. 2–12.

Yumoto, T. 1999 The objectives and protocols of IBOY 2001 (International Biodiversity Observation Year 2001). In: Chou, C.-H., Waller, G. R. & Reinhardt, C. (eds.) *Biodiversity and Allelopathy: From Organisms to Ecosystems in the Pacific*. Academia Sinica, Taipei, pp.29–35.

# Chapter editor

Dieter Mueller-Dombois, Ecosystem Division, PSA Task Force on Biodiversity, Hawai'i

# Index

ABTI 174
Acari 149
ahupua'a 195
algae 163
algal picoplankton (APP) 154
all-biota taxonomic inventory (ABTI) 173
amphibians (Amphibian) 93, 94
amphipod (Amphipoda) 116, 142
ant 45, 59, 75
arthropod (Arthropoda) 116, 137, 149
  communities 203
  fauna 42
  fauna of the bark surface 58
  free-living…fauna 53
  soil meso- 65
  soil micro- 65
ascidians 172
attached algae 149, 150
Australia 176, 177

bacteria 117, 120, 151, 152, 153, 155
Banda-Flores Seas 176
Bark spraying 58
baseline information 167
baseline studies 164
basic sampling unit 201
bathymetric profile 118
beating 74
Beaufort Sea 177
bedrock 119, 136, 137, 142
bee 52
beetle 59
belt transect method 83
benthic algae 150

benthos 114, 115, 118, 119, 148
Bering Sea 177
  West 166
biodiversity 163
  assessments 12
  functions 199
  level 12
Biodiversity Observation Year (BOY) 1
BIOMARE 174, 181
BioRap 169
bird 93, 94
bivalves (Bivalvia) 116, 137, 144
Blue Belt 183
BOD 120
bottom materials 136, 137, 139, 140, 141, 143, 144, 146, 148
bottom particles 139
bottom sample 139, 140, 141, 143, 144, 145
bottom samplers 137
bryozoan 163, 172
  mounds 173
buddy diving 122
butterfly bait trapping 72

Cambodia 176
Canada 177
canopy knockdown 53
capacity building 174
Caribbean 165
casting net 127
catch per unit effort 72
*Chaoborus* 151
China 177
chironomid 138, 145

# Index

chlorophyll a 172
ciliate 120, 153, 156
*Cladophora* 150
clear lake 114, 118, 156, 157
climate change 162
Cnidaria 181
cnidarian corals 171
coastal region 162
COD 120
Coleoptera 44, 48, 147
Collembola 59
CoML 174, 179, 181
comparative plot size 192
compensation depth 118
continental shelf 165
Convention on Biological Diversity 162
coral 165
  reef 171, 173, 178, 180
  hermatypic…reef 166
  soft 163
Coral Sea 176
core 166
  and satellite sites 30
crustaceans 147, 151
CTFS 204
cyprinid 133, 134
Cypriniform 133, 134

DAPI solution 153
database 122
decapod (Decapoda) 171, 178, 116, 181
decomposition 42
deep sea 165
Diptera 45, 48, 52
dissolved organic carbon (DOC) 120
dissolved oxygen (DO) 120, 123
DIVERSITAS 2, 177
DIWPA 2, 164, 165, 166, 167, 171, 172, 174, 176, 177, 178, 179
  Forest Protocol 201
  PABITRA Relationship 184
  intersects with 189
Drosophilidae 67
dynamic map 204
dynamics patterns 202

East China Sea 166, 177
echinoderms (Echinodermata) 171, 181
echinoids 178
echo-sounder 119
Ekman-Birge sampler 137, 138
electrical conductivity 120
electrofishing 124, 126
environmental variables 36
Ephemeroptera 147
ethanol 132, 133, 136, 148
Exclusive Economic Zone 162

FDP plots 204
Federated States of Micronesia 176
fence trapping 94
field data sheet 124, 129, 130
Fiji 176
filamentous green algae 150
filtration 153, 154, 155
fish 114, 115, 116, 124, 125, 126, 127, 128, 129, 131, 133, 134, 140, 163, 171,178
fixation 120, 124, 131, 132, 133, 139, 143, 146, 148, 151
fluorescent dye 153, 155
food-web 163
forest animals 201
formalin 131, 132, 133, 136, 143, 144, 145, 146, 147, 148, 152

211

formula 187

gastropod (Gastropoda) 116, 142, 165
Gastrotricha 148, 149
geoecology 190
gill net 124, 125, 129
GIS 173
Global 200 176
glutaraldehyde 153, 154
goby 129
gonadosomatic index 136
GPS 119, 120, 129
gradient 165
  environmental 190
  latitudinal 27, 165
gravel 119, 136, 140
Great Barrier Reef 166, 176
Green Belt 183
guild concept 202
guild spectrum 203
gum-chloral liquid 148, 149

haemocytometer 150, 156
hand collecting 78
hand net 126, 128, 129, 136, 137, 139, 140, 144, 145, 146, 147
Hawai'i 204
  IBP 195, 197, 198, 199
Hemiptera 48, 147
Heteroptera 45
heterotrophic nanoflagellates (HNF) 120, 151, 152, 155
holothurians 178
Homoptera 45, 52
honey baits 76
human impacts 164
hydroids 163, 172
Hydrozoa 148
Hymenoptera 45

hypotheses on species 199

IAVS 205
IBOY (International Biodiversity Observation Year) 1, 164, 166, 169, 170, 174, 177, 178, 179
  study plot 205
  aims of...2001 185
  goals for...2001 185
Indian Ocean 165
Indonesian 176
  Seas 166
insect (Insecta) 116, 143, 147, 151
Insular Pacific 166
inventory 4, 9, 29, 163, 164, 171
invertebrates 163
island biogeography theory 186
Isthmus of Kra 176
IUBS 177

Japan 166, 176, 177
  Society for the Promotion of Science 179

kelp 166
  forests 162
Kuroshio Current 166

label 129, 133
labelling 129, 132, 146, 148, 150, 151
Lake Baikal 157
Lake Biwa 114, 118, 120, 137, 157
Lake Khuvsgol 114
Lake Poso 114
Lake Taupo 114
Lake Toba 114
landscape 17
large macrofauna 181
leaf litter sampling 63, 78

Lepidoptera 44, 45, 69
library 181
light
  condition 118, 120
  traps 43
litter
  sifting 74
  -traps 39
littoral zone 114, 116, 117, 118, 119, 136, 141, 157
Long-Term Ecological Research 165
long-term monitoring 164, 167
loose (not submerged) stone 136, 139, 142, 143
Lord Howe Island 176
Lugol's solution 150, 153, 156

macroalgae 168, 169, 170, 171, 172, 178, 180, 181
macrobenthos 116, 136, 137, 139, 140, 141, 142, 144, 145, 146, 147, 150
macrofauna 169, 170, 174, 178
macrophyte 116, 117, 118, 119, 136, 137, 139, 141, 142, 144, 145, 146, 169, 170, 171, 178
macrozooplankton 151
main hypothesis 194
Malaise traps 48
Malaysia 165, 176
mammals 93, 97
  marine 172
mangrove 165, 171, 173, 176, 178, 180
mariculture 163
Marine
  BioRap 168
  station networks 167

marine
  protected areas (MPAs) 167
  reptiles 172
MARS 167, 181
  -BIOMARE 179
Mauna Loa transect 198
Megaloptera 143, 147
meiobenthic 116, 139, 148, 149
  net 139
meiobenthos 147, 148, 149
meiofauna 170, 174, 178
meiozoobenthos 139, 147, 148
mesh bag 139, 142
mesh size 125, 128, 129, 137, 139, 144, 151
mesozooplankton 151
meteorological data 36
methylcellulose 156
microbenthic 116
Micronesian 176
microphytoplankton 153
microplankton 152
minnow trap 126, 127
molecular studies 171
mollusks (Mollusca) 116, 137, 147, 171, 178, 181
monitoring 4, 9, 29, 163, 194
Mountain Transect Studies 195
mud 119, 136, 139, 140, 147
muddy bottom 128
muddy sand 140
multi-profiler 120
myriapods 59
mysis 151

NAML 181
nanozooplankton 151
Nansei Shoto 176
Nations Environment Programme 162

nematods (Nematoda) 148, 149
New Caledonia 176
New Guinea 176
New Zealand 166, 177, 178
  Shelf 166
nitrogen 120
Norfolk Island 176
North Korea 177
Northern Australian Shelf 166
nutrients 120

OBIS 179, 181
Odonata 143, 147
*Oedogonium* 150
oligochaetes (Oligochaeta) 138, 148, 149
opossum shrimp 151
organism groups 197
oxbow lake 114, 157, 158
Oyashio Current 166

PABITRA 185, 189
  concept 187
  transect sites 204
  website 206
Pacific area map outlining 188
Pacific Ocean 165
Palau 176
Papua New Guinea 176
part of minimal sampling kit 174
pebble bottom 117
pelagic zone 117
Perciform 133, 134
permanent plots 194
pH 120, 123, 172
phenology 42
Philippines 176
phosphorus 120

photographs 119, 131, 132, 143, 148, 149
phytoplankton 117, 152
  micro- 120, 151, 156
  nano- 120, 151, 156
  nano 153
  pico- 120
picoplankton 151, 152
pinnipeds 173
Pisces 181
pitfall 94
  traps 59
plankton 114, 115
  net 151, 152
planktonic organisms 150
plastic bag 140, 142, 144, 145, 146, 147
Plathelminthes 149
Plecoptera 143, 147
plot establishment 33
plot size 191
preservation 124, 131, 132, 143, 148
pressure-state-response model 163
primulin 155
pristine 180
pristiness 167
protected areas 18
Protista 148
protozoan 151
PSA Task Force on Biodiversity 206
Psocoptera 45

quadra-method 76
quadrat 119, 125, 136, 137, 140, 141, 142, 150, 169, 170

Rapid Biodiversity Assessment 164
recognizable but unnamed taxa 174

## Index

reference collection 171
reptiles 93, 94
research vessel (boat) 121, 145, 151
rock 119, 142
rocky
  bottom 117, 137
  shore 157
rotifer 151, 152
Rotifera 148, 149
RTUs 174
Russia 177, 178
Ryukyu Islands 166

Sabah 157
salinity 172
Samoa 176
sample 133
  lots 132
  storage 181
sampling lots 129, 146, 148, 150, 151
sampling protocol 199
sand 119, 136, 140, 147
sandy bottom 128, 148
satellite sites 166
scleractinian 163
SCOPE 177
screen net 125, 126, 129
scuba (SCUBA) 113, 114, 118, 121,
    122,123, 125, 129, 136, 137,
    139, 157, 158
sculpins 129
seabird 172, 173
seagrass 163, 168, 169, 170, 171,
    172, 173, 178, 180, 181
seamounts 163
Sea of Japan 166
  northern 177
Sea of Okhotsk 166, 177

seaweed 163
Secchi disk 118, 120
Seine net 128
shellfish beds 173
silicate 120
silt 119, 139, 147
silty bottom 137
Siluriform 134
Sloan Foundation 179
socio-economic benefits 20
soft bottom 117, 136, 137, 140, 142,
    143
soil samples 77
soil sampling 65
Solomon Islands 176
South China Sea 166
South Korea 177
species lots 132
species turnover 16
species/area curves 191, 193
spiders 59, 73
*Spirogyra* 150
sponge 118, 119, 172
  beds 173
standard walk 71
stone 119, 139
  sample 143, 145
  unit sampling 119, 142, 143, 145,
    150
stony bottom 125
substratum types 118, 119, 129,
    136
Sulawesi Sea 176
Sulu Sea 176
Sulu-Celebes Seas 166
Sundaland 176
surf zone 118
systematic expertise 174

tall-trees 38
Tardigrada 149
target organisms 7
Tasks for Vegetation Ecology in the PABITRA Net 205
temperature 170, 172
termites 80
  nests 84
Thailand 176
thermocline 118
three practical methods 191
Tonga 176
topographic map 120, 129
topography 136
toxic-algal blooms 162
transect
  counts 94
  line 118, 119, 120, 150
transparency 118, 120, 141
tree-count plots 201
trichopterous (Trichoptera) 45, 144, 147
Trizma-hydrochloride solution 155
Tropical Island Belt 183
tubeworm reefs 173
tubeworms 163
Tüllgren funnel 63, 65
Tundai 157
turbellarian 142
turbid lakes 114
turtle 173
Tuvalu 176
type of substratum 140

*Ulothrix* 150
unconsolidated bottom 136, 137, 139, 141, 142
under-stone samples 145
UNESCO 177

unionid 137
upland/lowland design 194
USA 177
UV-B radiation 162

Van-Dorn water sampler 152
Vanuatu 176
vascular plants 116
vegetation 187
  map 205
  of the Pacific Islands 187
  sample 141, 144, 145
vertebrate 92, 163
vertical
  profiles 120
  transects 194
video 119, 131, 141, 143
Vietnam 176
volcamic high island 196

water
  depth 117, 119, 123, 129, 130, 151
  sample 119, 152, 154, 155, 156
  sampler 151
  temperature 118, 120, 123, 130
wentworth convention 170
window traps 52

Yellow Sea 166, 177

zooplankton 117, 151, 152